T0321723

Growth, Cancer, and the Cell Cycle

The Molecular, Cellular, and Developmental Biology

Experimental Biology and Medicine

GROWTH, CANCER, AND THE CELL CYCLE

The Molecular, Cellular, and Developmental Biology

Edited by

PHILIP SKEHAN
and
SUSAN J. FRIEDMAN

The University of Calgary
Calgary, Alberta, Canada

Humana Press · Clifton, New Jersey

DEDICATION

This book is dedicated to the memory of Jacob Duerksen, a friend, colleague, and fellow member of The International Cell Cycle Society who contributed much to the organization of this conference. His untimely death greatly saddened those of us who knew him.

Library of Congress Cataloging in Publication Data

International Cell Cycle Conference (10th : 1984 : Banff
National Park)
Growth, cancer, and the cell cycle.

(Experimental biology and medicine)
Includes index.
1. Cells—Growth—Congresses. 2. Cell proliferation—
Congresses. 3. Cell cycle—Congresses. 4. Cancer cells—
Growth—Congresses. I. Skehan, Philip. II. Friedman,
Susan J. III. Title. IV. Series. [DNLM: 1. Cell Cycle.
2. Cell Transformation, Neoplastic. QZ 202 G884]
QH605.I525 1984   599'.031   84-22468
ISBN 0-89603-071-7

©1984 The Humana Press Inc.
Crescent Manor
PO Box 2148
Clifton, NJ 07015

Printed in the United States of America

PREFACE

Cell growth, one of the most fundamental of biological processes, has long been among the least understood. On April 24–28, 1984 scientists convened from around the world in Canada's Banff National Park for The International Cell Cycle Society's 10th Conference. Their purpose was to evaluate recent developments in the field of cell proliferation and to explore the interrelationship between cell growth, development, and differentiation, and proliferative diseases such as cancer. *Growth, Cancer, and the Cell Cycle* collects those conference papers that present the most recent advances in this field.

The first section of the book is Gene Expression and Development During Growth. It examines the structure and function of chromatin, DNA unwinding proteins, and nonhistone nuclear proteins, then explores transcriptional, translational, and post-translational regulation during the cell cycle and the interrelationship and coordinate regulation of cell growth, differentiation, and gene expression.

The second section, Growth Activation and Dormancy, focuses upon the events that occur during the transition between active cell growth and proliferative quiescence. The role of DNA strand breaks, protein kinase activity, growth regulatory factors, and the cytoskeleton are examined.

Section three discusses The Topology of the Cell Cycle. It reviews genetic approaches for determining the sequence of events and causality relationships that comprise and coordinate the many separate processes involved in cell cycle progression and describes the use of multiparameter flow cytometry to characterize the mammalian cell cycle and intracellular metabolic and transitional growth states.

The final section of the book, Neoplastic Transformation, explores the role of tumor cell heterogeneity in circulatory metastasis and of activator RNA in cellular phenotype transformation, presents the latest advances in the establishment of normal and neoplastic gastrointestinal cell lines, describes new methods for cancer diagnosis and of cell cycle deconvolution by image analysis, and critically evaluates the concept that cancer is a disease of abnormal cell growth.

Philip Skehan
Susan J. Friedman

CONTENTS

II. GROWTH ACTIVATION AND DORMANCY

III. TOPOLOGY OF THE CELL CYCLE

IV. NEOPLASTIC TRANSFORMATION

CONTRIBUTORS

S. ADÁMI · Cancer Research Group, Pilisborosjenö, Hungary

J. B. AUST · The University of Texas Health Science Center at San Antonio, San Antonio, Texas

RAMESH C. ADLAKHA · Department of Chemotherapy Research, The University of Texas M. D. Anderson Hospital and Tumor Institute, Houston, Texas

R. G. ALLEN · Department of Biology, Southern Methodist University, Dallas, Texas

H. ALTMANN · Institute of Biology, Research Center Seibersdorf, Austria

A. T. ANNUNZIATO · Department of Basic and Clinical Research, Scripps Clinic and Research Foundation, La Jolla, California

EDWARD BAPTIST · Department of Zoology, University of Georgia, Athens, Georgia

HÉLÈNA BIGO · Department of Genetics, The University of Texas M. D. Anderson Hospital and Tumor Institute, Houston, Texas

R. CURTIS BIRD · Department of Molecular Biology and Genetics, College of Biological Science, University of Guelph, Guelph, Ontario, Canada

N. L. R. BUCHER · Department of Pathology, Boston University School of Medicine, Boston, Massachusetts

G. C. CANDELAS · Department of Biology, University of Puerto Rico, Rio Piedras, Puerto Rico

T. M. CANDELAS · Department of Biology, University of Puerto Rico, Rio Piedras, Puerto Rico

G. CSABA · Department of Biology, Semmelweis University School of Medicine, Budapest, Hungary

ZBIGNIEW DARZYNKIEWICZ · Memorial Sloan-Kettering Cancer Center, New York, New York

DAVID T. DENHARDT · Cancer Research Laboratory, University of Western Ontario, London, Ontario, Canada

CATHARINE L. DEWAR · Oncology Research Group and Department of Pharmacology, The University of Calgary, Calgary, Alberta, Canada

P. DIXON · The University of Texas Health Science Center at San Antonio, San Antonio, Texas

DYLAN R. EDWARDS · Cancer Research Laboratory, University of Western Ontario, London, Ontario, Canada

D. ESCOBAR · The University of Texas Health Science Center at San Antonio, San Antonio, Texas

PETER A. FANTES · Department of Zoology, University of Edinburgh, Edinburgh, Scotland

SUSAN J. FRIEDMAN · Oncology Research Group and Department of Phamacology, The University of Calgary, Calgary, Alberta, Canada

HELENA GABOR · Children's Hospital of Northern California, Oakland, California.

WIEL GEILENKIRCHEN · Department of Zoology, State University, Utrecht, The Netherlands

I. GINZBURG · Department of Neurobiology, The Weizmann Institute of Science, Rehovot, Israel

W. L. GREER · Department of Biochemistry, University of Alberta, Edmonton, Alberta, Canada

MARGARET S. HALLECK · Molecular and Cell Biology Program, The Pennsylvania State University, University Park, Pennsylvania

S. K. HANKS · Department of Basic and Clinical Research, Scripps Clinic and Research Foundation, La Jolla, California

YEN-MING HSU · Department of Biochemistry, Michigan State University, East Lansing, Michigan

ANAND P. IYER · Microbiology Program, The Pennsylvania State University, University Park, Pennsylvania.

FRED A. JACOBS · Department of Molecular Biology and Genetics, College of Biological Science, University of Guelph, Guelph, Ontario, Canada

G. C. JOHNSTON · Department of Microbiology, Dalhousie University, Halifax, Nova Scotia, Canada

J. G. KAPLAN · Department of Biochemistry, University of Alberta, Edmonton, Alberta, Canada

A. KOVÁCS · Cancer Research Group, Pilisborosjenö, Hungary

P. KOVACS · Department of Biology, Semmelweis University School of Medicine, Budapest, Hungary

MARGARIDA KRAUSE · Department of Biology, University of New Brunswick, Fredericton, New Brunswick, Canada

JOLANTA KURZ · Department of Biology, University of New Brunswick, Fredericton, New Brunswick, Canada

KATHERINE LUMLEY-SAPANSKI · Molecular and Cell Biology Program, The Pennsylvania State University, University Park, Pennsylvania

ANDREA M. MASTRO · Microbiology Program, The Pennsylvania State University, University Park, Pennsylvania

J. McCARTHY · Department of Biology, Southern Methodist University, Dallas, Texas

J. A. McGOWAN · Children's Service, Shriners Burns Institute and Massachusetts General Hospital, Boston, Massachusetts

M. P. MOYER · The University of Texas Health Science Center at San Antonio, San Antonio, Texas

C. NATIONS · Department of Biology, Southern Methodist University, Dallas

G. NÉMETH · Cancer Research Group, Pilisborosjenö, Hungary

A. ORTIZ · Department of Biology, University of Puerto Rico, Rio Piedras, Puerto Rico

N. ORTIZ · Department of Biology, University of Puerto Rico, Rio Piedras, Puerto Rico

GORDHAN L. PATEL · Department of Zoology, University of Georgia, Athens, Georgia

SHARON A. PISHAK · Microbiology Program, The Pennsylvania State University, University Park, Pennsylvania

POTU N. RAO · Department of Chemistry Research, The University of Texas M. D. Anderson Hospital and Tumor Institute, Houston, Texas

SESHA REDDIGARI · Department of Zoology, University of Georgia, Athens, Georgia

JON A. REED · Molecular and Cell Biology Program, The Pennsylvania State University, University Park, Pennsylvania

O. M. RODRÍGUEZ · Department of Biology, University of Puerto Rico, Rio Piedras, Puerto Rico

LARRY ROSENBERG · Department of Chemotherapy Research, The University of Texas M. D. Anderson Hospital and Tumor Institute, Houston, Texas

W. E. RUSSELL · Children's Service, Shriners Burns Institute and Massachusetts General Hospital, Boston, Massachusetts

CHINTAMAN G. SAHASRABUDDHE · Department of Pathology, University of Texas M. D. Anderson Hospital and Tumor Institute, Houston, Texas

ROBERT A. SCHLEGEL · Molecular and Cell Biology Program, The Pennsylvania State University, University Park, Pennsylvania

R. L. SEALE · Department of Basic and Clinical Research, Scripps Clinic and Research Foundation, La Jolla, California

BRUCE H. SELLS · Department of Molecular Biology and Genetics, University of Guelph, Guelph, Ontario, Canada

ROSE SHEININ · Department of Microbiology, University of Toronto, Toronto, Canada

Z. SIMON · Cancer Research Group, Pilisborosjenö, Hungary

R. A. SINGER · Departments of Medicine and Biochemistry, Dalhousie University, Halifax, Nova Scotia, Canada

SANGRAM SISODIA · Department of Zoology, University of Georgia, Athens, Georgia

PHILIP SKEHAN · Oncology Research Group and Department of Pharmacology, The University of Calgary, Calgary, Alberta, Canada

ROBERT J. SKLAREW · New York University Research Service and School of Medicine, Goldwater Memorial Hospital, Roosevelt Island, New York

R. D. SMITH · Department of Basic and Clinical Research, Scripps Clinic and Research Foundation, La Jolla, California

MARION J. SNIEZEK · Microbiology Program, The Pennsylvania State University, University Park, Pennsylvania

UIK SOHN · Department of Biology, University of New Brunswick, Fredericton, New Brunswick, Canada

L. SZALAI · Cancer Research Group, Pilisborosjenö, Hungary

J. M. TAKÁCS · Cancer Research Group, Pilisoborosjenö, Hungary

JAMES THOMAS · Oncology Research Group and Department of Pharmacology, The University of Calgary, Calgary, Alberta, Canada

PETER, E. THOMPSON · Department of Zoology, University of Georgia, Athens, Georgia

O. TÖRÖK · Department of Biology, Semmelweis University School of Medicine, Budapest, Hungary

L. URBANCSEK · Cancer Research Group, Pilisborosjenö, Hungary

A. VÉRTESY · Cancer Research Group, Pilisborosjenö, Hungary

JOHN L. WANG · Department of Biochemistry, Michigan State Unviersity, East Lansing, Michigan

ROELAND van WIJK · Department of Molecular Cell Biology, State University, The Netherlands

KENNETH R. WILLIAMS · Department of Molecular Biophysics and Biochemistry, Yale University, New Haven, Connecticut

DAVID A. WRIGHT · Department of Genetics, The University of Texas M. D. Anderson Hospital and Tumor Institute, Houston, Texas

PAUL G. YOUNG · Department of Biology, Queen's University, Kingston, Ontario, Canada

A. M. ZIMMERMAN · Department of Zoology, University of Toronto, Toronto, Ontario, Canada

S. ZIMMERMAN · Division of Natural Science, Glendon College, York University, Toronto, Ontario, Canada

ACKNOWLEDGMENTS

We would like to thank Ms. Donna Wilson for her excellent assistance in organizing the Conference, and The Alberta Heritage Fund for Medical Research, Carl Zeiss, Inc., The National Science Foundation, and The University of Calgary for their financial support of the Conference.

SECTION 1

*GENE EXPRESSION AND DEVELOPMENT
DURING GROWTH*

STRUCTURES INHERENT TO CHROMATIN ACTIVE IN

TRANSCRIPTION AND REPLICATION

Annunziato, A. T., Smith, R. D., Hanks,
S. K., and Seale, R. L.

Department of Basic and Clinical Research
Scripps Clinic and Research Foundation
10666 North Torrey Pines Road
La Jolla, California 92037

INTRODUCTION

The structure of the nucleosome and higher order
transitions of the unit fiber of chromatin have been
described in considerable detail (1-3). Despite this
body of general structural knowledge, the function and
structure of regions actively engaged in replication and
transcription are just beginning to be understood (4-6).

Given certain similarities in the action of nucleic
acid copying enzymes, active transcription and
replication domains may have structural features in
common. In each case a multisubunit enzyme reads a
single strand of DNA in a linear, processive fashion.
DNA is denatured in both processes, although it is
localized to the site of enzyme binding in the case of
RNA polymerase, while denaturation is more extensive and
more accessory proteins are probably involved in DNA
replication. The template for both activities is not
free DNA, but chromatin. Thus, both RNA and DNA
polymerase molecules must either have the capacity to
read DNA complexed in nucleosomes, or the chromatin must
adopt a configuration so as to allow polymerases to
function. The overall effect may be quite similar, and
would seem to differ more in degree than in kind.

3

In our recent studies on the structure of newly replicated and of transcriptionally active chromatins, several parallels have become apparent. In this paper, we present highlights of these studies and stress the commonalities of the structural transitions between these two types of active chromatin.

CHROMATIN REPLICATION

The enhanced sensitivity of newly replicated chromatin to nucleases has been established for nearly ten years (7). However, the basis for this property is only now becoming clear. We presently understand this transient nuclease sensitivity to be due, at least in part, to a lag between the synthesis of DNA and the assembly of nucleosomes on 50% of the new DNA, in order to restore the histone complement. Thus, nascent chromatin consists of a nucleosomal (parentally derived) and a non-nucleosomal component in approximately equal proportions. In the subsequent discussion we consider the properties of each.

While DNA gels of digestion products clearly indicated that the nuclease-resistant component of new chromatin was nucleosomal (8-11), it was not clear from analysis of purified DNA whether these particles represented typical nucleosomes or structures that had been altered in composition, or configuration, and whether they were in an extended or folded higher order structure. In order to investigate these questions, composite agarose-polyacrylamide gels were utilized to separate nuclease digestion products as the nucleoprotein entities, core, HMG- and H1-nucleosomes (12, 13). Nuclei from HeLa cells incubated for 30 or 60 seconds in ^{3}H-thymidine were briefly digested with micrococcal nuclease (MNase) and then soluble chromatin was prepared two ways. In the first, total chromatin was prepared by lysis of nuclei in low ionic strength EDTA buffer, a step that solubilizes >80% of bulk chromatin (Fig. 1, lane T). Alternatively, chromatin was eluted in stepwise increments of NaCl concentration (0.1 to 0.6 M) in the presence of divalent cations. This procedure fractionates chromatin on the basis of molecular weight and protein composition (14). Monomers lacking H1 and enriched in HMG proteins are exclusively present in the

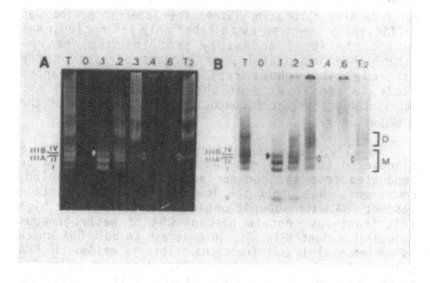

Figure 1. Distribution of newly replicated DNA in mononucleosomes and oligonucleosomes. HeLa cells were labeled for 30 sec. with [^{3}H]-thymidine. Chromatin was fractionated (text) and subjected to electrophoresis in agarose-polyacrylamide gels. Lane designations correspond to the NaCl molarity used for elution. Lanes T and T.2 contain chromatin released from nuclei with 2 mM EDTA without salt exposure (T), or after elution with 0.2 M NaCl (T.2). Nucleosomes are labeled in accord with ref. 12. A, ethidium bromide stain; B, fluorograph.

0.1 M eluate, while H1-monomers combined with nucleosome oligomers predominate in the 0.3 M eluate (Fig. 1A). An additional fraction was prepared as a control, in which monomers were largely removed by 0.2 M NaCl extraction, and all the remaining chromatin was released in a single step by 2 mM EDTA (Fig. 1A, lane T.2); this yields the soluble, high molecular weight chromatin in one fraction.

A fluorogram of pulse-labeled DNA (Fig. 1B) summarizes several important aspects of newly replicated chromatin (15). First, the average molecular weight of newly replicated chromatin is smaller than that of its bulk chromatin counterparts in the same fractions (Fig. 1, lanes T, 0.3, and T.2). Secondly, monomer nucleosome heterogeneity is essentially identical between bulk and pulse-labeled chromatin. Since the labeling period was sufficiently short so as to label only 4-8 nucleosomes behind the fork, accessory proteins must be re-established immediately, or perhaps persist on the particles throughout replication. Thus, enhanced nuclease sensitivity of these nucleosomes is not attributable to radical compositional alterations of the unit particle.

Finally, using this fractionation protocol, we discovered that the nucleosomal component could be separated from the non-nucleosomal, or unassembled, component. Note that the low salt fractions contain nascent DNA with subunit organization, while the high salt fractions contain nascent DNA of heterogeneous molecular weight (Fig. 2), in contrast to bulk DNA which is nucleosomal in all fractions. This is evident in DNA gels, rather than in particle gels, since high molecular weight nucleoprotein does not resolve well in composite gels. That the smeared patterns of newly replicated chromatin represent replication intermediates was demonstrated by labeling cells for 15 minutes in order for the mature chromatin pattern to predominate; in this case the subunit pattern in the fluorograph exactly superimposed that of the ethidium stain (15).

The amount of DNA comprising the two components (nucleosomal and non-nucleosomal) of newly replicated chromatin was approximately equivalent. While 90% of bulk material is eluted at 0.1 - 0.3 M NaCl, only 50% of

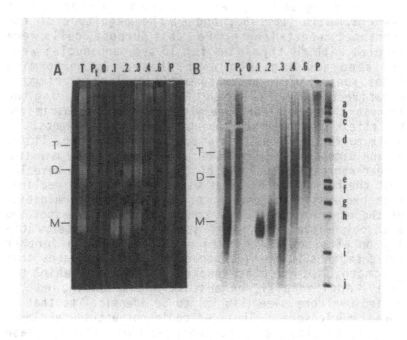

Figure 2.   DNA subunit character of newly replicated chromatin.   DNA was prepared from chromatin prepared as in Figure 1 and subjected to electrophoresis in an SDS-polyacrylamide gel.   Lanes O to P are labeled as in Figure 1.   Lane P_t contains the DNA of the EDTA-insoluble fraction.   M, D and T indicate nucleosome monomers, dimers and trimers.   A, ethidium bromide stain; B, fluorograph.

the pulse-labeled chromatin is (15). The remaining 50%
of newly replicated DNA is eluted only at high ionic
strength (0.4-0.6 M NaCl), and   10% is refractory to
solubilization (compared to 1% of bulk).

   The fractionation of the two constituents of
chromatin replication intermediates has enabled us to
determine their various properties. It was noted in
several earlier studies that the nucleosome repeat-length
immediately behind the fork was shortened (9, 16, 17).
We sought to determine 1) whether this was simply due to
more rapid degradation, hence shortened oligomer size of
new chromatin (18, 19), and 2) the magnitude of the
shortened repeat-length. For this purpose, cells were
labeled with ^{3}H-thymidine for 30 sec. and nuclei were
isolated and digested with MNase. By performing
digestions at 0° in order to minimize linker trimming
relative to linker cleavage (20), and by measuring the
largest oligomer clearly resolved, we doubly minimized
the effect of linker shortening on size measurements. In
order to further monitor for possible nucleosome sliding
during digestion, points were taken from 1 to 60 minutes.
Linear regression analysis of the data (20, 21) revealed
that the bulk histone repeat was 186-188 bp, and declined
with progressive digestion to 174 bp at the termination
of the experiment (22). In contrast, subunits associated
with pulse-labeled DNA were substantially shorter, 165-
167 bp after only 1 minute of digestion. No further
decrease in subunit size occurred. This indicates that
chromatosomes are close-packed immediately behind the
fork. Additionally, exhaustive digestion showed the
nucleosome core size (146 bp) to be identical to that of
normal nucleosomes. Thus, we could not attribute close-
packing to sliding of H1-depleted cores during MNase
digestion or to partially unravelled nucleosomes (22).

   Next, the properties of chromatin replication in
cycloheximide were examined. In cycloheximide, the dual
properties (i.e., nucleosomal and non-nucleosomal DNA) of
newly replicated chromatin were found to persist. When
nucleosomal heterogeneity was examined as in Fig. 1, a
full complement of accessory proteins, H1 and HMGs were
evident. Interestingly, when incubation in cycloheximide
was prolonged to 20 minutes or longer, the abbreviated
spacing of the nucleosomal component matured to the bulk

repeat-length, while the unassembled component continued to accumulate, requiring concomitant protein synthesis for assembly and maturation (22).

The nuclease sensitivity of nascent chromatin is maintained in cycloheximide due primarily to exposure of non-nucleosomal DNA. This was evident in the overall nuclease sensitivity of the DNA labeled in the presence of cycloheximide. Upon examination of the nucleosomal component, it was clear that it was more sensitive to nucleases than bulk nucleosomes. Therefore, it seemed possible that part of nuclease sensitivity could also be due to other factors, e.g., a maximally decondensed unit fibril. Histone hyperacetylation has been extensively correlated with extended chromatin and with transcriptionally active chromatin (23-26), but definitive experimental confirmation of these correlations has yet to appear. In order to test the possible involvement of histone acetylation in replication, cells were briefly labeled for 30 secs. to 30 mins. with ^{3}H-thymidine in the presence of the deacetylase inhibitor, sodium butyrate (27). In periods longer than 10 minutes, nucleosome assembly has occurred, and residual nuclease sensitivity due to factors other than unassembled DNA becomes detectable. Approximately 50% of the nuclease sensitivity is retained, even after 30 minutes. Upon removal of sodium butyrate, the nuclease sensitivity of the labeled regions reverted to that of bulk chromatin (28). We attribute the persistence of nuclease sensitivity of nascent nucleosomes due to the maintenance of a relatively extended conformation of the region, caused or potentiated by histone N-terminal acetylation. Electron microscopic studies of acetylated chromatin have subsequently confirmed this prediction (29).

In the next set of experiments we examined the structure of the non-nucleosomal component of newly replicated chromatin. By first eluting the nucleosomal component of MNase-digested nuclei in moderate ionic strength buffer, properties of the heterogeneous, i.e., non-nucleosomal nucleoprotein could be examined. It was apparent (e.g., see Figure 1) that the non-nucleosomal component was not free DNA. Naked DNA is degraded 40-fold faster than bulk chromatin under these conditions.

The average molecular weight of the non-nucleosomal DNA was actually greater than that of the pulse-labeled nucleosomal DNA, at all stages of digestion. Secondly, after nuclease cleavage into relatively small (e.g., 400-800 bp) fragments, heterogeneous DNA was not eluted at physiological ionic strength, but required subsequent exposure to elevated ionic strength $\geq$ 0.3 M NaCl for release, in contrast to nascent nucleosomes.

In order to determine whether any underlying nucleoprotein structure was associated with the heterogeneous material, it was redigested with MNase in a controlled fashion so as to cleave and trim connecting DNA interspersed with putative nuclease-resistant complexes (Figure 3). This experiment was performed either with MNase or with Hae III for initial cleavage; the result was the same in both cases. Secondary digestion of the heterogeneous DNP unmasked nucleosomes imbedded within unusually long stretches of DNA (Figure 3). The amount of nucleosomal protection of this component was on the order of 40-60%, representing a significant fraction of newly replicated chromatin.

This result is inconsistent with models in which parental nucleosomes follow only one arm of the replication fork. In consideration of several models of parental nucleosome segregation (30), the data best fit a mechanism whereby nucleosomes are distributed to both arms of the replication fork in clusters. The smeared component of nascent chromatin results from nucleolytic cleavages within the heterogeneous material, and also from one cleavage in a heterogeneous domain and one within a nucleosomal domain, yielding nucleosomes with extra-long terminal linker DNA. These terminal linkers are of heterogeneous size; after trimming, discrete nucleoprotein particles are revealed. The heterogeneous linker DNA is associated with insoluble nuclear structures, causing not only its insolubility, but also the insolubility of nucleosomes to which it is attached (30).

In summary, we have determined that newly replicated chromatin has two components. One is nucleosomal, exists in a relatively extended (i.e., lacks higher order structure) conformation, and has a decreased repeat-

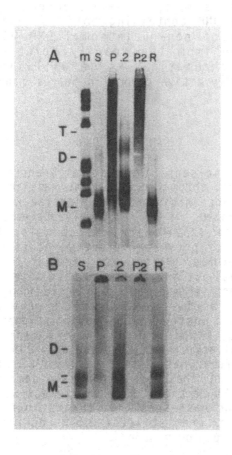

Figure 3. Sequential digestion of nascent heterogeneous nucleoprotein. HeLa cells were labeled with [^{3}H]-thymidine for 30 min. in the presence of cycloheximide. Isolated nuclei were digested with MNase and soluble chromatin released with 2 mM EDTA (lane 5). The insoluble pellet (lane P) was extracted with 0.2 mM NaCl to elute residual nucleosomes (lane .2). The remaining pellet (lane P.2) was redigested with MNase, lane R. Lane M, marker fragments. Panel A, DNA gel; Panel B, composite nucleoprotein gel. Both are fluorographs.

length. The second component is a non-nucleosomal DNA-protein complex that alternates on the same segment of newly replicated DNA with nucleosome clusters. Nucleolytic fragmentation of newly replicated chromatin yields nucleosomes, non-nucleosomal DNP, and heterogeneous DNP containing both components. The insolubility of non-nucleosomal DNP allows its fractionation from the readily soluble nascent nucleosomal chromatin. As we will show below, transcriptionally active chromatin shares many of these properties.

TRANSCRIPTION

The nuclease-sensitive state of transcribed genes relative to bulk chromatin was first made in chicken β-globin chromatin; in tissues in which the β-globin gene is not transcribed, it is relatively nuclease resistant (31). Nuclease sensitivity was soon found to be a property also of potentially active genes as well as quiescent genes that have had a history of transcriptional activity. It was then determined that not only the active gene, but also surrounding chromatin of a larger active domain were equally DNase I sensitive. For instance, the ovalbumin gene and two related genes, X and Y, which have widely differing transcriptional activities, comprise only 20% of a 100 kb domain of uniform DNase I sensitivity (32). In chicken globin chromatin, the DNase I-sensitive domain has been extended >1 kb to the 5' side of the active gene; the extent of 3' nuclease-sensitive chromatin is presently unknown, but extends at least 3.5 kb downstream, and includes the β-globin gene (33).

We sought to examine these parameters in mouse globin chromatin, and to relate them to known features of newly replicated chromatin. The mouse globin genes lie in a linked domain with the embryonic and β-homologous genes at one end, and the adult β-major and β-minor genes at the other (34) (Figure 4). In mouse erythroleukemia (MEL) cells, the adult β-globin genes are potentially active, and transcription can be induced by exposure of the cells to agents that stimulate erythroid differentiation (35).

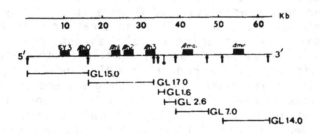

Figure 4.   Map of the β-globin gene family in MEL cells.
Cloned restriction fragments are designated GL followed
by their lengths in kilobase pairs.   R1 (⬆) and Bam H1
(●).

Thus, we compared the nuclease-sensitivities of regions spanning the globin gene domain in non-induced and in hexamethylene-bisacetamide (HMBA)-induced cells (36). The large probes, GL 7.0 and GL 14.0 (Figure 4), containing the β-major and β-minor genes, respectively, were significantly more DNase I-sensitive than sequences in the inactive, embryonic and β-homologous, globin gene chromatin. The DNase I-sensitive domain extends upstream of the β-major gene approximately 7 kbp, where a rather distinct boundary (< 1 kb) separates it from the resistant domain (37).

When these results were compared with those obtained with a different endonuclease, micrococcal nuclease (MNase), a similar pattern emerged, but there were significant differences (Figure 5). First, while differential sensitivities between the active and inactive domains occurred at the same boundary observed with DNase I, the overall nuclease sensitivity of the active domain became more marked in the induced cells. Secondly, the sensitivity of the active domain was not constant; the regions homologous to the β-globin probes GL 7.0 and GL 14.0 were significantly more sensitive than the GL 2.6 5' flanking sequence that lies immediately upstream of the β-major gene but within the active globin domain.

In order to examine the non-uniform sensitivity in more detail, a series of smaller probes across the β-major domain were utilized (Figure 6). The pre-induced, potentially active, domain had a rather uniform MNase sensitivity, including the 5' flanking, coding, and 3' flanking sequence (Figure 6). While an increased MNase sensitivity occurred with HMBA induction, it was significantly greater in the coding sequence than in the 5' flanking DNA. In fact, the nuclease sensitivity was not strictly confined to the coding sequence, but extended into the 3' flanking region for approximately 1 kb (37), and corresponded to the primary transcription unit (48, 58).

The subunit repeat-length of pre-induced, and HMBA-induced β-globin genes was then measured in order to test chromatin rearrangement associated with gene activity (38). The nuclease repeat-length was increased,

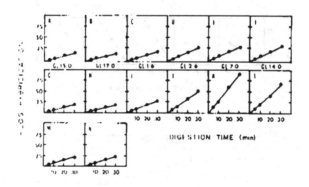

Figure 5. MNase sensitivity of sequences within the β-globin gene family. Following digestion of nuclei, DNA was isolated and hybridized to the indicated sequences. Lanes A-F, uninduced cells, lanes G-L, induced cells. As a control, cloned probe to the non-expressed immunoglobulin gene Cμ 12 was hybridized to uninduced (M) and induced (N) DNA, as above.

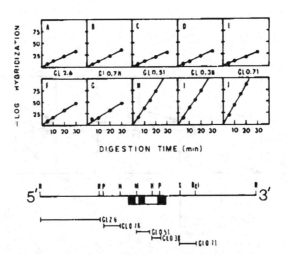

Figure 6.   MNase sensitivity of specific sequences near
the β-major globin gene. The labeled probes GL 2.6, GL
0.78, GL 0.51, GL 0.38, and GL 0.71 were hybridized to
MNase digested DNA from uninduced (A-E) or induced (F-J)
MEL cell nuclei.   Indicated restriction sites are R,
EcoRI; P, PstI; H, HindIII; M, MboI; X, XbaI; and Bgl,
BglII.

interestingly, in the active globin domain of both uninduced and induced cells by about 11 bp, and remained unchanged in the inactive domain (38). In cells that do not express globin the repeat length in both regions of the globin domain was the same as that of bulk chromatin.

We next sought to relate the differential nuclease sensitivities to properties of the unit nucleosome fiber, versus possible higher order structures. For this purpose, we utilized the property of chromatin to reversibly adopt an extended or a compacted configuration by manipulation of the ionic milieu. At very low ionic strength, and in the absence of divalent cations, the higher order folding interactions are lost, generating extended chromatin. Upon restoration of physiological ionic strength, or divalent cations, chromatin recondenses (40, 41).

The criterion of the extended-folded conformational transition was applied to the mouse globin domain. Under normal ionic strength conditions, a large differential in DNase I sensitivity exists between the inactive (GL 15.0) domain and the active (GL 7.0) domain. When divalent cations are removed, the inactive GL 15.0 domain acquires nuclease sensitivity equivalent to that of the active GL 7.0 domain (39). However, the sensitivity of the active domain is unchanged by these conditions. These results suggest that the inactive domain is nuclease resistant relative to the active domain due to compaction of nucleosomes into higher order chromatin solenoidal or superbead configurations (40, 41) of inactive regions, and an extended conformation of active regions. Hence, the active domain already exists in an extended conformation and is not affected by conditions that unravel higher order structures.

The converse experiment is to measure the extent of refolding of chromatin decondensed at low ionic strength. This was performed by restoration of NaCl to 40 mM (partial refolding conditions), or to 100 mM or $MgCl_2$ to 3 mM (maximal refolding conditions). In response to restored ionic strength, the inactive GL 15.0 domain recovered about 90% of its original nuclease resistance while there was no measurable effect on the nuclease sensitivity of the active GL 7.0 domain (Fig. 7). In an

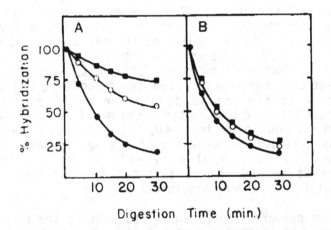

Figure 7. Effect of decondensation-recondensation transitions of chromatin on DNase I sensitivity. EDTA swollen nuclei were digested directly after addition of 50 μM MgCl$_2$ (●), or after addition of 40 mM NaCl (O) or 100 mM NaCl (■). Panel A, inactive GL 15.0 domain; Panel B, active GL 7.0 domain.

attempt to determine the basis for this effect, histone
H1 was removed at pH 4.5 and the domains tested for
nuclease sensitivity (39). H1 removal had the same
effect as reduction in ionic strength; the inactive
domain became sensitive, and the active domain was not
affected by this parameter.

Another genetic locus amenable to study of its
active and dormant states is the heat shock 70 kd (hsp
70) protein gene of Drosophila. Unlike the mouse β-
globin genes, hsp 70 is both DNase I and MNase resistant
before induction, and the sensitivity to both nucleases
is dramatically increased in response to heat shock,
concomitant with a loss of well-defined nucleosomal
structure (42, 44) (Figure 8). In this experiment,
instead of preparing total DNA from MNase-digested
nuclei, soluble chromatin (75% DNA) was first eluted in
2 mM EDTA, and the insoluble nuclear pellet DNA (20%)
was also prepared. The ethidium stained patterns of
chromatin from both control and heat-shock nuclei show
the typical nucleosome ladder. When hybridized to
plasmid pPW232.1 and pPW229.1, containing the hsp 70
coding sequence, the hsp 70 gene was nuclease-resistant,
relative to bulk chromatin, as previously shown (41).
Upon induction by heat shock, not only was there an
increase in nuclease sensitivity and perturbation of the
nucleosome ladder, but also there was a 4-fold shift of
coding sequences into the nuclear pellet fraction (Figure
8). Furthermore, the structure of hsp 70 chromatin in
the soluble and insoluble fractions was different;
nucleosomal structures were eluted in 2 mM EDTA, while
the smeared, apparently non-nucleosomal component was
enriched in the pellet. Upon redigestion of the smeared
pellet material, nucleosomes were apparent in 165 bp-
resistant nucleoprotein complexes (Figure 8). This
result bears a strong similarity to that found in the
fractionation of newly replicated chromatin into
nucleosomal and insoluble, heterogeneous nucleoprotein
embedded with nucleosomes.

To further define this phenomenon, probes to the 5'
and 3' flanking regions of hsp 70 were utilized (45).
The 5' flanking DNA does not undergo a loss of nucleosome
ladder, and exhibits a less dramatic shift in solubility
in response to heat-shock. In contrast, a probe located

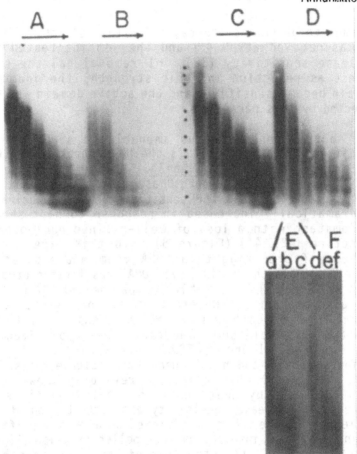

Figure 8. Nucleosomal structure as a function of expression of the hsp 70 gene of Drosophila Kc cells. Nuclei were isolated from control or from heat-shocked nuclei, digested with MNase for the indicated periods, and lysed with 2 mM EDTA. DNA was prepared from the soluble fraction (S2) and the pellet (P), subjected to electrophoresis in agarose gels and blot-hybridized to cloned DNA from the hsp 70 coding sequence (text). Panels A and C are control nuclei S2 and P, respectively. Panels B and D are heat-shock S2 and P, respectively. Digestion times are 1, 2, 4, 8, 16, and 32 min. in each panel Panels E, F. Redigestion of heat-shock pellet nucleoprotein. Lane a, residual 1-minute nucleoprotein redigested for 4 min. (lane b), or 32 min. (lane c). Lane d, residual 2-minute nucleoprotein redigested for min. (lane e) or 32 min. (lane f).

downstream from the 3' end portrayed a distinct loss of structure, and shift in solubility characteristics in parallel with the coding sequence as in Figure 8. Although the measurement here is different from the digestion parameters measured for the β-major globin gene, the parallel is striking; activation affects chromatin structure of not only the coding sequence, but considerable 3' flanking sequence, while 5' flanking chromatin is not affected.

DISCUSSION

In investigations of apparently unrelated active chromatin structures, transcriptionally active and newly replicated chromatin, we have found a strong commonality in structural transitions. Not only are both forms of active chromatin nuclease-sensitive, but also this property can be ascribed, at least in part, to an open, more accessible higher order structure of chromatin in both regions. The transcriptionally active chromatin domain was extended, and in addition, it was unable to fold under conditions that condense bulk, H1-containing chromatin (39). In newly replicated chromatin, we conclude that blockage of histone deacetylation was sufficient to prevent full higher order structure formation (29), while nucleosome assembly was not inhibited (28).

In the case of transcriptionally active chromatin, the nuclease-sensitive, decondensed state was not confined to the coding sequence, but extended both in the 5' and 3' directions. In the mouse globin gene domain, the 5' boundary of nuclease sensitivity was mapped to a rather precise boundary, 7 kb upstream from the β-major gene (37). Although DNAse I sensitivity of the active domain was uniform in coding and flanking regions, this was not the case with MNase. MNase recognizes different structural features of chromatin, and was a more sensitive indicator of gene activity, versus potential. The coding region became even more sensitive to MNase upon transcription than other regions of the active domain, and this sensitivity extended $\geq$ 1 kb into the 3' flanking region (39).

Increased nuclease sensitivity of transcription units has been correlated with loss of the canonical nucleosome ladder (42-44). When integrity of the nucleosome ladder was examined in both types of chromatin, newly replicated and transcriptionally active, there was a distinct smearing of the nucleosome ladder. When the smeared material was subsequently re-digested, cryptic nucleosomes were revealed, due to the more rapid nuclease action on linkers than on the relatively nuclease-resistant histone-DNA complex (30, 45).

Disruption of the nucleosome ladder of the transcribed globin and hsp 70 genes was not confined to the coding sequences, but extended into 3' chromatin, in parallel with the nuclease-sensitivity (37, 43, 45). Accompanying the loss of typical chromatin structure of both newly replicated and transcriptionally active chromatins is a shift in the solubility properties. The heterogeneous material is decidedly more insoluble than the typical nucleosomal component, allowing its fractionation. Among the possibilities for this solubility change are association with a distinct nuclear matrix upon which transcription and replication occur (46, 47), or simply the relatively insoluble nature of the myriad of activities involved in DNA copying. Since the sole parameter is insolubility, both arguments would presently appear to hold equal weight.

Disruption of the nucleosome ladder and shift in insolubility are clearly associated with the traverse of polymerases. Note that in transcribed regions, the 5' flanking DNA does not participate in these changes, while the 3' flanking chromatin does. It will be of considerable interest to map the 3' boundary of transcription in relation to the boundary of chromatin perturbation, since transcription can continue for extensive distances past the 3' coding terminus (48).

Disruption of the nucleosome ladder in active chromatin might be ascribed to either nucleosome unfolding, or to temporary dislodgement of the histone octamer. If the nucleosome has undergone a conformational change that is responsible for smeared MNase patterns, then one would predict that the pattern would remain smeared, irrespective of the extent of digestion. We observed, to the contrary, that the heterogeneous nucleoprotein isolated from both newly replicated and transcriptionally active chromatin contains within it nuclease-resistant nucleosomes typical of bulk particles. Thus, we conclude that active chromatin has _fewer_ nucleosomes per unit length of DNA. Electron microscopic visualization of transcription units support this contention (55 -57). Passage of either RNA or DNA polymerase may be concomitant with histone octamer dissociation in front of, and reassociation at some distance behind the polymerase. Nucleolytic fragmentation of this region would yield chromatin fragments with irregular, extra-long linkers that exhibit no apparent structure in partial digests, and nucleosomes in extended digests.

A prediction from this model that is amenable to test is that the extent of disruption of the nucleosome ladder is a function of the frequency of transcription initiation. Infrequently transcribed genes should appear nucleosomal, while disruption of the nucleosome ladder is maximal for genes with high transcriptional activity.

The long-standing observation that nucleosomes are associated with actively transcribed regions (31, 49), then, cannot be taken as evidence that nucleosomes, _per se_, are transcribed. If the histone octamer is not permanently associated with a given DNA sequence, then octamers must mix neighbors as a result of both transcription and replication. A formidable body of evidence has gathered in recent years to support this contention (51-54).

REFERENCES

1. McGhee, J. D., and Felsenfeld, G. (1980). Ann. Rev. Biochem. _49_, 1115-1156.
2. Cartwright, I. L., Abmayr, S. M., Fleischmann, G., Lowenhaupt, K., Elgin, S. C. R., Keene, M., and Howard, G. C. (1983). CRC Critical Rev.: Biochem. _13_, 1-86.
3. Igo-Kemenes, T., Horz, W., and Zachau, H. G. (1982). Ann. Rev. Biochem. _51_, 89-121.
4. Mathis, D., Oudet, P., and Chambon, P. (1980). Prog. Nucleic Acids Res. Mol. Biol. _24_, 2-49.

5. De Pamphilis, M. L., and Wassarman, P. M. (1980).
    Ann. Rev. Biochem. 49, 627-666.
6. Annunziato, A. T., and Seale, R. L. (1983). Mol.
    and Cellular Biochem. 55, 99-112.
7. Seale, R. L. (1975). Nature 255, 247-249.
8. Seale, R. L. (1976). Cell 9, 423-429.
9. Seale, R. L. (1978). Proc. Nat. Acad. Sci. USA 75,
    2717-2721.
10. Weintraub, H. (1976). Cell 9, 419-422.
11. Hildebrand, C. E., and Walters, R. A. (1976).
    Biochem. Biophys. Res. Commun. 73, 157-163.
12. Todd, R. D., and Garrard W. T. (1977). J. Biol.
    Chem. 252, 4729-4738.
13. Bakayev, V. V., Bakayeva, T. C., and Varshavsky, A.
    J. (1977). Cell 11, 619-629.
14. Sanders, M. M. (1978). J. Cell Biol. 79, 97-109.
15. Annunziato, A. T., Schindler, R. K., Thomas, C. A.
    Jr., and Seale, R. L. (1981). J. Biol. Chem. 256,
    11,880-11,886.
16. Levy, A., and Jakob, K. M. (1978). Cell 14, 259-
    267.
17. Murphy, R. F., Wallace, R. B., and Bonner S. (1978).
    Proc. Natl. Acad. Sci. USA 75, 5903-5907.
18. Lohr, D., Corden. J., Tatchell, K., Kovacic, R. T.,
    and Van Holde, K. E. (1977). Proc. Natl. Acad. Sci.
    USA 74, 79-83.
19. Jackson, V., Marshall, S., and Chalkley, R. (1981).
    Nucleic Acids Res. 9, 4536-4581.
20. Noll, M., and Kornberg, R. D. (1977). J. Mol. Biol.
    109, 393-404.
21. Sperling, L., Tardiev, A., and Weiss, M. C. (1980).
    Proc. Natl. Acad. Sci. USA 77, 2716-2720.
22. Annunziato, A. T., and Seale, R. L. (1982).
    Biochemistry 21, 5431-5438.
23. Simpson, R. T. (1978). Cell 13, 691-699.
24. Vidali, G., Boffa, L. C., Bradbury, E. M., and
    Allfrey, V. C. (1978). Proc. Natl. Acad. Sci. USA
    75, 2239-2243.
25. Mathis, D. J., Oudet, P., Wasylyk, B., and Chambon,
    P. (1978). Nucleic Acids Res. 5, 3523-3547.
26. Nelson, D. A., Perry, M., Sealy, L., and Chalkley,
    R. (1978). Biochem. Biophys. Res. Commun. 82, 1346-
    1353.
27. Riggs, M. G., Whittaker, R. G., Neuman, J. R., and
    Ingram, V. M. (1977). Nature 268, 462-464.

28. Annunziato, A. T., and Seale, R. L. (1983). J. Biol. Chem. 258, 12675-12684.
29. Frado, L. L., Annunziato, A. T., Seale, R. L., and Woodcock, C. L. F., submitted.
30. Annunziato, A. T., and Seale, R. L., submitted.
31. Weintraub, H., and Groudine, M. (1976). Science 193, 848-856.
32. Lawson, G. M., Knoll, B. J., March, C. J., Woo, S., Tsai, M.-J., and O'Malley, B. W. (1982). J. Biol. Chem. 257, 1501-1507.
33. Wood, W. I., and Felsenfeld, G. (1982). J. Biol. Chem. 257, 7730-7736.
34. Jahn, C. L., Hutchinson, C. A. III, Phillips, S. J., Weaver, S., Haigwood, N. L., Voliva, C. F., and Edgell, M. H. (1980). Cell 21, 159-168.
35. Nudel, U., Salmon, J., Eitan, F., Terada, M., Rifkind, R., Marks, P.A., and Banks, A. (1977). Cell 12, 463-469.
36. Osborne, H. B., Bakke, A. C., and Yu, J. (1982). Cancer Res. 42, 513-518.
37. Smith, R. D., Yu, J., and Seale, R. L. (1984). Biochemistry 23, 785-790.
38. Smith, R. D., Seale, R. L., and Yu, J. (1983). Proc. Natl. Acad. Sci. USA 80, 5505-5509.
39. Smith, R. D., Yu, J., and Seale, R. L. (1984). Biochemistry, in press.
40. Thoma, F., Koller, Th., and Klug, A. (1979). J. Cell Biol. 83, 403-427.
41. Zentgraf, H., Muller, U., and Franke, W. W. (1980). Eur. J. Cell Biol. 23, 171-188.
42. Wu, C., Bingham, P. M., Livak, K. J., Holmgren, R., and Elgin, S. C. R. (1979). Cell 16, 797-806.
43. Levy, A., and Noll, A. (1981). Nature 289, 198-203.
44. Levinger, L., and Varshavsky, A. (1982). Cell 28, 375-385.
45. Hanks, S. K., and Seale, R. L., submitted.
46. Robinson, S. I., Nelkin, B. D., and Vogelstein, B. (1982). Cell 28, 99-106.
47. Vogelstein, B., Pardoll, D. M., and Coffey, D. S (1980). Cell 22, 79-85.
48. Hofer, E., and Darnell, J. E., Jr. (1981). Cell 23, 585-593.
49. Garel, A., and Axel, R. (1976). Proc. Natl. Acad. Sci. USA 73, 3966-3970.

50. Annunziato, A. T., and Seale, R. L. (1983). Mol.
     Cell. Biochem. 55, 99-112.
51. Annunziato, A. T., Schindler, R. K., Riggs, M. G.,
     and Seale, R. L. (1982).
52. Jackson, V., and Chalkley, R. (1981). J. Biol.
     Chem. 256, 5095-5103.
53. Russev, G., and Hancock, R. (1982). Proc. Natl.
     Acad. Sci. USA 79, 3143-3147.
54. Fowler, E., Farb, R., and El-Saidy, S. (1982).
     Nucleic Acids Res. 9, 4563-4581.
55. McKnight, S. L., Sullivan, N. L., and Miller, O. L.
     Jr. (1976). Prog. Nucleic Acid Res. Mol. Biol. 19,
     313-318.
56. Lamb, M. M., and Daneholt, B. (1979). Cell 17, 835-
     848.
57. Mathis, D., Oudet, P., and Chambon, P. (1980).
     Prog. Nucleic Acids Res. Mol. Biol. 24, 1-55.
58. Salditt-Georgieff, M., Sheffery, M., Krauter, K.,
     Darnell, J. E. Jr., Rifkind, R., and Marks, P. A.
     (1984). J. Mol. Biol. 72, 437-450.

CHROMATIN CHANGES IN SKELETAL MUSCLE CELLS DURING THE FUSION TO MYOTUBES

O. TÖRÖK, H. ALTMANN[*]

Department of Biology, Semmelweis University of Medicine, Budapest, Hungary
[*] Institute of Biology, Research Centre Seibersdorf, Austria

ABSTRACT

Differentiation of cells is connected to changes in chromatin structure. Skeletal muscle cells were either labeled with ^{3}H-thymidine, to estimate DNA synthesis and NAD^+ was used as the precursor for poly(ADP-ribose)-synthesis in permeabilized cells. Parallel the fusion of myoblasts to myotubes were observed, different times of the cultivation period. At the time when myoblast fusion was high, DNA-synthesis was low, but poly(ADP-ribose)-synthesis was increasing. Autoradiographic data confirmed the results on semiconservative DNA-synthesis. Methoxybenzamide at very low concentrations stimulates poly(ADP-ribose)-synthesis and also myoblast fusion, at higher concentrations both processes are inhibited.Pulse chase experiments using a 5 min ^{3}H-thymidine pulse showed that most of the specific radioactivity is located first in the micrococcus nuclease sensitive region, possibly as Okazaki units, which are elongated during the chase period and were found at that time also in the micrococcus nuclease resistant region. There are only small differences in spacer/core relationship at different times of myogenesis. During the most active fusion period newly synthesized poly(ADP-ribose) is located more in the MNase sensitive part of chromatin. At that time more DNA

27

strandbreaks could be detected by nucleoidsedimentation technique.

INTRODUCTION

A relationship exists between inhibition of the
cell cycle and differentiation of certain cells
(1, 2). Culture conditions under which DNA-synthesis in the cells is inhibited, for example high
local density or dense monolayer, are conditions
for the induction of differentiation. The most
important mechanism in muscle cell differentiation
is the fusion of myoblasts into multi nucleated
myotubes. Cell surface lectine receptors and
lipids of the plasma membrane play important roles
in the process of myoblast fusion. This process is
preceded and accompanied by an increase in membrane
fluidity (3). In the time period from 24 to 72
hours after cultivation of cells is the interval
during which the cells have ceased to proliferate and undergo a burst of fusion activity (4).
There is evidence that synchronized cells blocked
in the S-phase show structural changes in chromatin (5). Chromatin structure seems to be influenced
by posttranslational modifications of histone and
nonhistone proteins. ADP-ribosylation and the
binding of poly(ADP-ribose) (PAR) to proteins have
an important role in regulation of the cell cycle,
but especially on DNA metabolism. PAR is synthesized from NAD by an enzyme attached to chromatin. In vitro experiments have shown, if PAR-
polymerase was incubated with chromatin, a relax-
ation of the native structure of chromatin was
observed and examined by electronmicroscopy (6).
PAR-polymerase activity is necessary for the
differentiation process (7). The role of ADP-
ribosylation in differentiation was first suggested
by Caplan and Rosenberg (8). They first found a
dependence of NAD level whether differentiated meso--
dermal cells of embryonic chicklimbs will express
myogenic or chondrogenic properties. Low levels of
NAD are correlated with chondrogenic expression
while inhibiting myogenic expression, while high
levels of NAD are correlated with myogenic expression and with inhibition of chondrogenic expression. Differentiation can also be reversibly

inhibited by nicotinamide starvation and the
lowering of the cellular NAD content of the myoblasts (7). There are controversial reports on the
formation of DNA strandbreaks during differentiation, but it seems possible that breaking and
rejoining is regulated by PAR-polymerase because
the activity of this enzyme is also important in
the process of gene rearrangement. Recently it
could be shown that topoisomerase I activity is
regulated also by ADP-ribosylation (9). Topoiso-
merases are enzymes which can catalyse the conser-
ted breaking and rejoining of DNA phosphodiester
bonds. The enzyme relaxes supercoiled DNA in the
absence of added cofactor and at least one part
can be coisolated together with PAR-polymerase.
During induction of differentiation in Friends
cells the specific activity from incorporated NAD
remained unaltered in nonhistone proteins (10).
Several authors also believe that PAR-polymerase
might primarily regulate the proliferative activity of cells undergoing differentiation. However,
PAR is also related to the regulation of expression
of different gene products during differentiation.
Also the average chain length of PAR was found to
be reduced during differentiation. The total content of nonhistone proteins was reduced in chromatin to about 50% by treatment with PAR inducers
(11). To study chromatin structure during differentiation,micrococcus nuclease digestion can be
used for characterisation of spacer and core
region in chromatin. At least 4 types of chromatin
structure can be distinguished by digestion kinetic
experiments. Significant differences in the digestion behaviour of chromatin from metaphase and
interphase have been detected by this method.
Furthermore the structure of newly replicated DNA
in S-phase differs from the bulk, in that it is
more easily degradated to acid soluble products by
micrococcus nuclease (12). In the development of
skeletal muscle, myogenic cells fuse to form
multinucleated mytobes which later develop into
mature muscle fibers. This transition from proliferating mygenic cells to multinucleated myotubes
occurs in cultures of dissociated mygenic cells
derived from newborn rats and such cultures provide a useful tool to study terminal differen-

tiation and changes in the chromatin during the
fusion to myotubes.

MATERIALS AND METHODS

Cells for culture were isolated from the
pectoral and legmuscle of newborn rats. The muscle
fragments were incubated $\emptyset$ Ca, Mg, containing PBS
for 30 min at 37°C, followed by trypsin digestion
(0.25% in Ca, Mg PBS for 30-45 min at 37°C). After
the incubation the free cells were collected in
complete medium (TC 199 + 10% foetal calf serum),
filtered through Nytal to remove debry, and collected by centrifugation. After discarding the
supernatant the cells were suspended in complete
medium. The cell suspension was divided in 3 ml
per tube in which coverglasses were put previously.
The cultures were maintained at 37°C for 10 days.
One group of cultures was labeled after 24 hours
of incubation by adding ³H-thymidine in serum free
TC 199 medium (0.5µCi per ml) for 60 min at 37°C.
At the end of labeling period cultures were washed
twice with TC 199 and cultivated further for
additional 72 hours till 10 days in nonradioactive
medium to determine the time of the prefusion cells
to myotube, in complete medium. Cultures were
fixed for autoradiography at 1, 4, 24, 30, 48 and
72 h. Cells on coverglasses were rinsed 4 times in
PBS, fixed in methanol 5 min, washed in destilled
water, airdried and coated in emulsion (Ilford G5).
The coverglasses were exposed in sealed boxes for
96 h at 4°C. After developing the autoradiographs
the cultures were stained with GIEMSA solution
(0.5%), and scored for the percent of mononucleated
myocytes and myotubes containing labeled nuclei. In
parallel we observed in vivo the differentiation
events from mononucleated myocytes to multinucleated myotubes by microcinematography too. At
different times of the cultivation, cells were
fixed and stained with May-Grünwald GIEMSA each
day, to estimate the fusion process. We scored
the number of myocytes and myotubes during this
period. A cell was considered fused if two or more
nuclei were clearly seen in a shared cytoplasm.
For each determination 10^3 nuclei were scored in
random fields. The ability of cells, expressing the

myogenic program to form multinucleated myotubes,
was examined as a marker for the differentiated
state. An increase in the number of nuclei per cell
indicates that myocytes are spontaneously fusing
to form multinucleated myotubes.

For the PAR-polymerase inhibition studies
methoxybenzamide (MBA) was used in various concentrations. In the first experiment 24 h after
the begining of the cultivation, cells were
treated and kept with methoxybenzamide (100µM)
containing medium for 5 days. In the second experiment only on the first of 5 days, methoxybenzamide was applied for 24 h to the cells. In the
third experiment different concentrations of
methoxybenazamide (10, 50, 100, 250 and 500µM and
1, 5 and 10 mM concentrations, during 5 days, beginning 24 h after cultivation) were used.

For determination of the PAR-synthesis, cells
were made permeable to NAD by hypoosmotic cold
shock (4°C for 15 min). The incubation buffer
(0.5 ml) consists of 0.01 M Tris, 0.25 M sucrose,
0.001 M EDTA, 0.03 M DTE and 0.004 M $MgCl_2$ pH 7.8.
After the incubation time at 4°C 1µCi $NAD^2$was
applied in 0.2 ml buffer (0.1 M Tris, 0.12 M $MgCl_2$
pH 7.8).The NAD was adenine labeled at 2, 8 ^{3}H,
3 Ci per mM NEN. The incubation time with labeled
NAD was 30 min at 37°C. The reaction was stopped
by cold PCA and radioactivity was counted in the
precipitate. For the determination of radioactivity within chromatin (^{3}H-thymidine incorporated in
DNA) crude chromatin was precipitated from cells
and micrococcus nuclease digestion was done. The
crude chromatin was isolated from cells through
lysis in a mixture of 0.5% Triton X 100, 0.1 M
EDTA, 2 mM Tris, pH 7.8 (according to 13). DNA
from chromatin was digested by 30 iU micrococcus
nuclease per sample for 10 min at 37°C. The radioactivity of supernatant and precipitate was
measured in a liquid scintillation counter. For
the determination of the DNA synthesis ^{3}H-thymidine
(0.1µCi/ml; 50-80 Ci(mmole NEN) was incorporated
for 1 h at 37°C.

Nucleoid sedimentation was done according to

Cook and Brazell (14). The cells are lysed in
Triton X 100, 2 M salt and a concentration of
chelating agents, sufficient to completely inhibit
nucleases. Such nucleoids contain naked histon free
DNA in supercoiled form within a cage of protein
and RNA. The centrifugation time of nucleoids in
the sucrose gradient was 60 min at 15.000 U/min
in a Ti 40 rotor of a Beckman ultracentrifuge.

RESULTS AND DISCUSSION

In Figure 1 the schedule of the rearrangement
experiments of chromatin is shown.

It was first found that rearrangements of chromatin structure took place following DNA repair
synthesis in eukaryotic cells after damage of DNA
(15). Similar processes have been shown to occur
in replicating DNA in both cellular and viral
systems (16, 17, 18, 19). Replicating chromosomal
DNA is digested by micrococcus nuclease more
rapidly and to a greater extent as DNA in nonre-
plicating chromatin, releasing small (3-7 S) nascent DNA fragments that are subsequently digested
completely (20). This group also found that pulse
labeled intact cells or nuclear extract was
digested about 5 fold faster and about 25% more

Figure 1

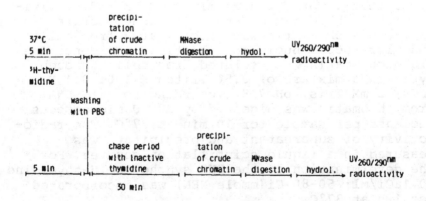

PULSE-CHASE EXPERIMENT

extensively, than uniformly labeled DNA in mature
viral chromosomes. Pulse chase experiments in vitro
revealed a time dependant chromatin maturation
process that involves two steps. First the conversion of prenucleosomal DNA into immature nucleosomal oligomeres and second maturation of newly
assembled chromatin into a structure with increased nuclease resistance. In figure 2 we see
the rearrangement of chromatin structure during
the differentiation steps of newborn rat muscle
cells in culture. When newborn rat muscle cells
are cultured intensive DNA synthesis starts
immediately, but on the second till the third day,
at the begining of myotube assembling, semiconservative DNA synthesis is markedly reduced. In contrast to human cells chromatin from rodant cells
is only to a small amount (10-20%) micrococcus
nuclease sensitive. The specific activity of
^{3}H-thymidine after a pulse of 5 min is much higher
in the micrococcus nuclease sensitive region
immediately after the pulse label. After a chase
period with inactive thymidine also the core
region becomes more and more radioactive. The
higher values of radioactivity comes from the fact
that ^{3}H-thymidine was still present in the pool of
nucleotides and during the chase period this amount
of ^{3}H-thymidine was also additionally incorporated
in DNA. In the three day old culture there was a

Figure 2

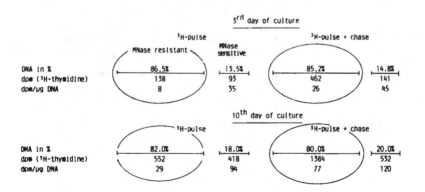

REARRANGEMENTS OF CHROMATINSTRUCTURE DURING DIFFERENTATION-STEPS
OF NEW BORN RAT SKELETAL MUSCLE CELLS IN CULTURE

lot of fusion of myoblasts to myotubes. In contrast
in the 10 day culture, at least under our conditions, there were only little fusion to myoblasts
visible. There was also again an increase in semi-
conservative DNA synthesis, the micrococcus
nuclease sensitive region becomes a little greater
and after a chase time of 30 min also the shift
from the micrococcus nuclease sensitive to the
resistant region was more pronounced. In figure 3
the ratio of the specific activity between micrococcus nuclease sensitive and resistant region is
shown. On the third day of culture the ratio
directly after the pulse labeling was 4.38 and
after the chase time 1.73, in the 10 day old
culture the ratio after the pulse was 3.24 and
after the chase time 1.56. There are generally
two theories which could explain the shift of the
activity of the micrococcus nuclease sensitive to
the resistant region. The first one would be a
continuous   changing of corehistones in their
position along the DNA. This process was discussed
as sliding of nucleosomes, and in the case of chromatin without damage of DNA, it is called con-
stitutive rearrangement. The other explanation
could be that there is for a limited time a
loosening in the core structure and therefore a

Figure 3

RATIO OF SPECIFIC ACTIVITY (dpm/µg) DNA OF $\frac{\text{MNase resistant}}{\text{MNase sensitive}}$

REGION IN CHROMATIN OF NEW BORN RAT SKELETAL MUSCLE CELLS IN CULTURE

	³H-thymidine pulse	³H-thymidine pulse + chase
3rd day of culture	4.38	1.73
10th day of culture	3.24	1.56

change   in this two micronuclease degradable
regions. In our described experiment the spacer
to core region remains constant or nearly constant
during the experimental time. An assambling of
Okazaki units occur during the chase period and
therefore also more activity is shifted to the
core region. To what extent the loosening process
is involved, in which possibly poly(ADP-ribose)
take part, cannot be distinguished up till now.
Relatively little is known about the posttranslational modification of chromatin proteins during
myogenesis. During the first proliferation stage
histones, especially H_2B, were phosphorylated. In
prefusion postmytotic cell phosphorylation of
histones H_2A, H_3 and H_4 declined, whereas all
histones decreased modification at the myotube
stages (21).

Some recent experiments have shown that phosphorylation of histone and nonhistone proteins
maybe are regulated by ADP-ribosylation. Changes
in the phosphorylation of histones during myogenesis reflect possibly withdrawal from the mytotic
cycle. In our experiments with poly(ADP-ribose)
distribution in chromatin of newborn rat muscle
cells we could find that always the micrococcus
nuclease sensitive region was highly active compared to the resistant region. But during the stage
of myotube fusion the activity was about doubly
high as in the stage of seven day old cultures.

The minimum   incorporation of ^{3}H-thymidine
during semiconservative DNA synthesis correlates
good with the presence of multinuclei cells
(figure 4).

In an parallel experiment, when DNA synthesis
was low, PAR synthesis had a maximum and precede
the myoblast fusion (figure 5).

Nucleoidsedimentation studies showed some DNA
strandbreaks in 3 and 7 day old cultures (figure 6).

F. Farzanek et al. (22) have demonstrated that
during the differentiation of primary chick
skeletal muscle cells in culture, DNA single strand

Figure 4

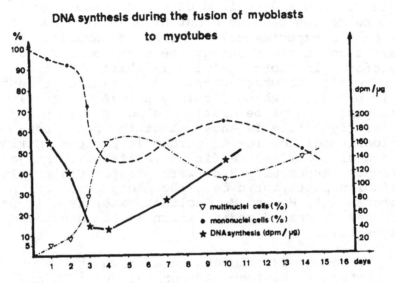

DNA synthesis during the fusion of myoblasts to myotubes

Figure 5

Poly(ADP ribose)synthesis during myogenesis

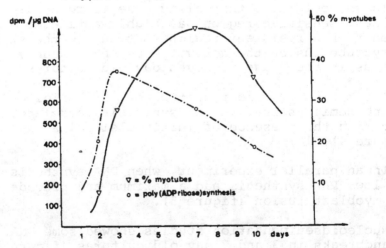

Figure 6

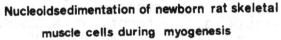

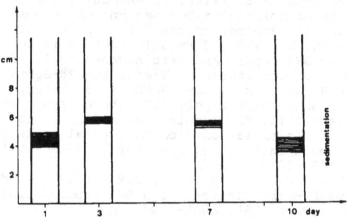

breaks appear. DNA breaks activate the PAR-polymerase, which correlate good to the obtained data
on PAR-synthesis. DNA breaking and rejoining,
regulated by PAR-polymerase, may be involved in a
general mechanism of differentiation.

REFERENCES

1. NISHIMUNE, Y., A. KUME, Y. OGISO and A.
    MATSUSHIRO, 1983, Induction of teratocarcinoma
    cell differentiation effect of the inhibitors
    of DNA synthesis. Exp.Cell Res. 146, 439-444.
2. CLAYCOMB, W.C., 1975, Biochemical aspects of
    cardiac muscle differentiation. J.Biol.Chem.
    250, 3229-3235.
3. PRIVES, J., M. SHINITZKY, 1977, Increased
    membrane fluidity precedes fusion of muscle
    cells. Nature 268, 761-763.
4. HERMAN, B.A., S.M. FERNANDEZ, 1982, Dynamics
    and topographical distribution of surface
    glycoproteins during myoblast fusion: A reson-
    ance energy transfer study. Biochemistry 21,
    3275-3283.
5. D'ANNA, J.D., D.A. PRENTICE, 1983, Chromatin
    structural changes in synchronized cells

blocked in early S phase by sequential use of
isoleucine deprivation and hydroxyurea blockade.
Biochemistry 22, 5631-5640.
6. JONGSTRA-BILEN, J., G. MURCIA, C. NIEDERGANG,
    G. POIRSER, M.E. ITTEL, P. MANDEL, 1982, The
    effect of poly(ADP-ribolisation) on chromatin
    structure. Biology of the Cell,p 45 Abstract
7. ALTMANN, H., 1983, Chromatin factors influ-
    encing DNA repair and carcinogenesis. 13.Int.
    Congr. of Chemotherapy, Vienna Aug.28-Sept.2.
8. CAPLAN, A.I., M.J. ROSENBERG, 1975, Interre-
    lationship between poly(ADP-ribose)-synthese,
    intracellular NAD levels, and muscle or carti-
    lage differentiation from mesodermal cells of
    embryonic chick limb. Proc.Natl.Acad.Sci.USA
    72, 1852-1857.
9. FERRO, A.M., N.P. HIGGINS, B.M. OLIVERA, 1983,
    Poly(ADP-ribosylation) of a DNA topoisomerase.
    J.Biol.Chem. 258, 6000-6003.
10. ZLATANOVA, J.S., P. SWETLY, 1980, Poly(ADP-
    ribosylation) of nuclear proteins in differ-
    entiating Fried cells. Biochem.Biophys.Res.
    Comm. 92, 1110-1116.
11. MORICKA, K., K. TANAKA, T. ONO, 1980, Poly(ADP-
    ribose) and differentiation of Friend erythro-
    leukemia cells. J.Biochem. 88, 517-524.
12. JALOUZET, R., D. BRIANE, H.H. OHLENBUSCH, M.L.
    Wilhelm, F.X. WILHELM, 1980, Kinetics of
    nuclease digestion of physarium polycephalum
    nuclei at different stages of the cell cycle.
    Eur.J.Biochem. 104, 423-431.
13. ALTMANN, H., I. DOLEJS, A. TOPALOGLOU, A. SOOKI-
    TOTH, 1979, Faktoren, die die DNA Reparatur in
    "Spacer and Core DNA" von Chromatin menschlicher
    Zellen beeinflussen. studia biophys. 76, 195-
    203.
14. COOK, P.R., I.A. BRAZELL, 1976, Detection and
    repair of single strand breaks in nuclear DNA.
    Nature 263, 679-682.
15. SMERDON, M.J., M.W. LIEBERMAN, 1978, Nucleosome
    rearrangemnt in human chromatin during UV-in-
    duced DNA repair synthesis. Proc.Nat.Acad.Sci.
    USA 75, 4238-4241.
16. DOLEJS, I., H. ALTMANN, 1980, Studies on dis-
    tribution of ^{3}H-thymidine incorporated into
    micrococcus nuclease sensitive and resistant

region in chromatin of chinese hamster ovary
(CHO) cells. IRCS Med.Sci. 8, 392-393.

17. HILDEBRAND, C.E., R.A. WALTERS, 1976, Rapid
assembly of newly synthesized DNA into chromatin subunits prior to joining of small DNA
replication intermediates. Biochem.Biophys.Res.
Comm. 73, 157-163.

18. SEALE, R.L., 1975, Assembly of DNA and protein
during replication in HeLa cells. Nature 255,
247-249.

19. KLEMPNAUER, K.H., E. FANNING, B. OTTO, R.
KNIPPERS, 1980, Maturation of newly replicated
chromatin of simian virus 40 and its host cell.
J.Mol.Biol. 136, 359-374.

20. CUSICK, M.E., T.M. HERMAN, M.L. DePAMPHILIS,
P.M. WASSARMAN, 1981, Structure of chromatin
at deoxyribonucleic acid replication forks:
Prenucleosomal deoxyribonucleic acid is rapidly
excised from replicating simian virus 40 chromosomes by micrococcal nuclease. Biochemistry
20, 6648-6658.

21. LOUGH, J., 1983, Declining histone phosphorylation during myogenesis revealed by in vivo
and in vitro labeling. Exp.Cell Res. 148, 437-
447.

22. FARZANEK, F., R. ZALIN, D. BRILL, S. SHALL,
1983, Involvement of DNA strand breaks and ADP-
ribosyl transferase activity in muscle cell
differentiation. Abstr.Europ.Workshop on ADP-
ribosylation of proteins, Berlin, p. 9.

A RAT LIVER HELIX-DESTABILIZING PROTEIN: PROPERTIES AND HOMOLOGY TO LDH-5

G.L. Patel[*], S. Reddigari[*], K.R. Williams[#],
E. Baptist[*], P.E. Thompson[*] and S. Sisodia[*]

[*]Department of Zoology, University of
Georgia, Athens, GA.

[#]Department of Molecular Biophysics and
Biochemistry, Yale University, New Haven, CT.

Retrieval of chromosomal information must entail various conformational modulations of the DNA duplex. Such changes may result from direct actions of regulatory proteins on DNA itself and/or indirectly via modifications of DNA-associated proteins. Proteins that affect conformational alterations in DNA have been identified in prokaryotes and eukaryotes (reviewed in ref. 1), and these changes are important in gene activity (reviewed in ref. 2). For example, the base pairing mechanism underlying sequence fidelity, either for replication or for transcription of DNA, requires that hydrogen bonds holding the two complementary strands of the DNA duplex be destabilized for strand separation. Accordingly, for some time we have been investigating DNA-binding proteins that exhibit such activity in vitro. We describe here a single-stranded DNA-binding protein purified from rat liver with characteristic DNA helix-destabilizing activity, and present surprising evidence which indicates extensive structural and functional homology between this rat liver helix-destabilizing protein (HDP) and lactic dehydrogenase-5 (LDH-5) from various sources.

Purification and Properties of an HDP from Rat Liver

Preferential binding to denatured DNA is a common characteristic of all HDPs (3). Therefore, we adopted a

41

differential DNA-affinity chromatography method (4) as a
first step to isolate single-strand specific DNA-binding
proteins from liver extracts. Either fresh or frozen
livers excised from Sprague-Dawley male rats (200-250 g)
were used as the source of protein. Typically, about
30-40 g of liver tissue pulp was homogenized in a Sorvall
Omnimixer with 300 ml of 20 mM Tris-HCl, pH 7.5, 50 mM
NaCl, 1 mM disodium EDTA, 0.1 mM DTT and 5% glycerol
(Buffer A, ref. 4) freshly supplemented with phenyl-
methylsulfonyl fluoride and dithiothreitol to final
concentrations of 0.1 and 1 mM respectively. The homo-
genate was processed as described (4) to obtain crude
liver protein extract, which then was fractionated by
affinity chromatography on native double-stranded and
denatured single-stranded DNA-cellulose columns prepared
according to Alberts and Herrick (5) with calf thymus DNA
(Sigma Chemical Co.) and Whatman CF11 cellulose powder.
Each column contained 1-1.5 g DNA in a bed volume of
200 ml. The extract was applied to columns connected in
tandem such that proteins not binding to double stranded
DNA immediately passed onto the single stranded DNA
column. After washing with the starting buffer the
columns were disconnected and individually eluted with
Buffer A containing 0.5 mg/ml dextran sulfate to remove
nonspecifically adsorbed proteins (4) and then with
Buffer A containing 2 M NaCl to obtain dextran sulfate
resistant DNA-binding proteins. SDS-polyacrylamide gel
electrophoresis patterns of the various chromatographic
fractions showed complex but distinct patterns of proteins which bound to native vs denatured DNA (data not
shown).

Proteins which passed through double stranded DNA-
cellulose but bound to single stranded-DNA, and which
were subsequently eluted with 2 M NaCl-Buffer A, then
were fractionated further by molecular sieve chromatography on Ultragel AcA 44 (LKB, Inc.). Chromatographic
fractions were assayed for helix-destabilizing activity.
A protein, which eluted at a position corresponding to
MW $\cong$ 120,000, melted poly [d(A-T)·d(A-T)] when mixed in
a suitable low ionic strength buffer (Buffer B: 5 mM
Tris-HCl, pH 8, 1 mM disodium, 0.1 mM DTT and 10%
glycerol, ref. 4) at room temperature. Molecular purity
of this fraction is shown by the single SDS-PAGE band
migrating as a protein of ~30,000 daltons (figure 1).

Figure 1.   SDS-polyacrylamide gel patterns of HDP purified from normal (a) and regenerating (b) livers of rat.

Routinely about 0.3-0.4 mg of this pure HDP is recovered
from the extract of 1 g liver.   Some physico-chemical
properties of this protein are summarized in Table 1.

Table 1.   Some Physico-chemical Properties of Rat Liver
HDP

Molecular Structure:	Tetramer of identical sub-units of ~30,000 M.W.
Stokes radius:	42.5 Å [a]
Frictional coefficient:	1.33 [b]
Amino acid composition:	Acidics/Basics = 1.63
pI:	> 9
$E^{1\%}_{280\ nm}$:	13.9 [c]

(a) Determined by gel filtration on sephadex G-150 with
catalase, yeast alcohol dehydrogenase, bovine serum
albumin and hemoglobin as standards.

(b) Calculated according to Siegel and Monty (13).

(c) Determined according to Babul and Stellwagen (14).

Helix-Destabilizing Activity of Rat HDP

We have studied the activity of rat HDP by moni-
toring its ability (a) to induce hyperchromicity in
selected synthetic polynucleotides at 25° C (isothermal
denaturation) and (b) to lower the thermal denaturation
temperature (Tm depression) of duplex natural DNA.   In
each case 1 cm light path cuvettes were employed and
hyperchromicity was monitored at 260 nm.

For isothermal denaturation assay, two alternative
methods were employed.   In one, the test DNA and HDP,
each diluted in Buffer B at 25° C, first were adjusted to
identical absorbance at 260 nm.   Then successive 0.15 ml
aliquots of the HDP were added to 1.5 ml of the DNA and
the absorbance recorded after each addition.   The alternate method employed a special teflon stoppered cuvette
with two equal compartments in tandem created by a
partial quartz partition (Hellma).   This cuvette allowed
measurement of the pre-interaction absorbance of equal
volumes of HDP and DNA placed on either side of the
partition.   The cuvette was oriented so that light passed
perpendicular to the partition through both compartments.
The cuvette next was inverted repeatedly to allow the HDP
and DNA to mix over the partition, and the post-
interaction absorbance was measured.   In both of these
methods, any alterations in absorbance upon mixing must
reflect a consequence of productive HDP-DNA interactions,
while non-productive mixing should exhibit no such
changes.

Studies on HDP-induced alterations in thermal denaturation temperature of DNA employed a Beckman ACTA III
double-beam spectrophotometer equipped with a temperature
regulated sample compartment, automatic sample changer,
temperature measurement via a probe placed directly in
one cuvette, and a recorder.   The absorbances of control
DNA and DNA-HDP samples were recorded as a function of
continuously changing temperature (ca. 1°/min) to generate denaturation and renaturation profiles.

Using the isothermal assay we have observed that the
liver HDP readily denatures poly [d(A-T)·d(A-T)] upon
mixing.   The extent of HDP-induced hyperchromicity
depends upon the ratio of HDP/DNA (w/w); a maximal

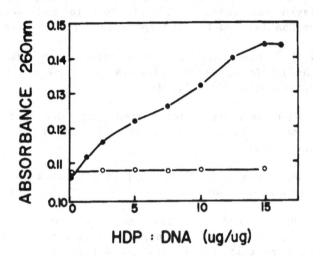

Figure 2. Effect of HDP/DNA (μg/μg) on melting of poly [d(A-T)·d(A-T)] and poly [d(G-C)·d(G-C)] in Buffer B. HDP and polynucleotides in Buffer B were adjusted so that each gave $A_{260 \, nm}$ = 0.107; reference cuvette contained Buffer B. To cuvettes containing 1.5 ml of poly [d(A-T)· d(A-T)], (●) or poly [d(G-C)·d(G-C)], (o), sequential aliquots of HDP were added and any change in $A_{260 \, nm}$ recorded.

hyperchromicity of 34% is obtained at HDP/poly [d(A-T)· d(A-T)] = 15 (Fig. 2). Trivial explanations of the observed increase in A_{260} such as turbidity due to precipitation or nucleolytic degradation are ruled out by the following:

(a) The assay reactions do not exhibit any increase in A_{314} nm over a wide range of the added protein, or changes in any other region of the absorption spectrum of the mixtures, except for specific increases in the absorption maximum around A_{260} nm.

(b) The purified HDP lacks nuclease activity when assayed for release of acid-soluble radioactivity from DNA labeled to high specific activity with ^{32}P.

(c) The hyperchromity resulting from HDP:poly [d(A-T)·d(A-T)] interaction is reversed readily by addition of NaCl, KCl, etc.

(d) When psoralen cross-linked poly [d(A-T)·d(A-T)] is used as a substrate, the HDP fails to induce hyperchromicity.

(d) When poly [d(G-C)·d(G-C)] is used as a substrate, no change in absorbance of mixture is observed.

Our HDP also destabilizes natural duplex DNAs, as reflected in their reduced thermal denaturation temperature. We have tested DNA from six different sources and with (G+C) contents ranging from 25-72%. In all cases a Tm depression of about 20° was observed (Fig. 3). Preliminary studies with monomeric and dimeric nucleosomes obtained by micrococcal nuclease digestion of liver chromatin showed that the HDP reduced their Tm dramatically by more than 40°C; however, because HDP treated nucleosomes showed some precipitation during the process of thermal denaturation, definitive interpretation of these data must await further studies. Nevertheless, the suggested effect of HDP on chromatin is relevant since in eukaryotes DNA is not naked and is associated with proteins.

Identification of the "Active-Site" of HDP

We have exploited the rapid isothermal assay for activity of HDP on poly [d(A-T)·d(A-T)] (described above) to explore the effects of chemical modifications of amino acid side-chains on the structure and activity of this protein. Since tyrosine residues have been shown to be involved in the interaction of many prokaryotic proteins with DNA (6), we initiated this aspect of our active-site study with modifications of this amino acid (7).

The rat HDP contains six tyrosines per subunit. We treated the protein with tetranitromethane (TNM) (8) and

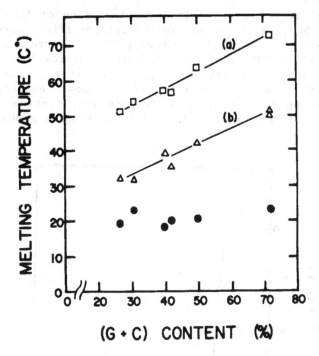

Figure 3. Effect of [G+C] content of natural DNA on HDP-induced depression of Tm. Pairs of samples were prepared to contain 5 μg DNA or 5 μg DNA + 50 μg normal liver HDP in a final volume of 1 ml Buffer. The Tm of these pairs were determined and plotted vs known [G+C] content of the DNA. (□) Tm of DNA control, and (△) Tm of DNA + HDP; (●) represent HDP-induced Tm depression. The natural DNAs employed in this experiment were: mouse satellite, C. perfringens, calf thymus, mouse main band, E. coli, and M. lysodeikticus.

KI$_3$ (9) to nitrate and iodinate, respectively, the tyrosine residues; we then measured consequent changes in its structure and activity. Evidence for modification of tyrosines was obtained by spectrophotometric measurements since the 3-nitrotyrosine product of the nitration reaction has absorbance maxima at 428 nm and 381 nm. Amino acid analysis confirmed that the other possible targets (tryptophan, methionine and histidine) of TNM were not affected. Cysteine, another target for TNM, was

shown not to be involved in the interaction of this HDP
with DNA because use of the sulfhydryl specific reagents,
N-ethylmaleimide and 5,5'-dithiobis (2-nitrobenzoic acid)
did not alter HDP activity.   Tyrosine modifications also
did not appear to introduce gross structural changes.
For example, SDS-PAGE of nitrated HDP showed absence of
crosslinking of subunits, a feature complicating some
cases of reactions with TNM.   Gel filtration on Sephacryl

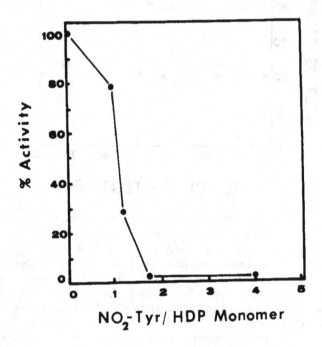

NO$_2$-Tyr/ HDP Monomer

Figure 4.   Effect of nitration of HDP on its activity.
HDP was treated with varying amounts of TNM to nitrate
tyrosine residues and assayed as described.   Hyperchromicity produced at 260 nm when 5 μg poly [d(A-T)·d(A-T)]
was mixed with 100 μg modified HDP was measured as
described.   Unmodified HDP control produced 28% hyperchromicity; this is expressed as 100%.   The number of
tyrosine residues modified as calculated from $A_{381\ nm}$.

S-200 and cross-reaction with anti-HDP rabbit serum also showed that tyrosine modifications did not elicit any gross conformational alterations of the protein.

However, tyrosine modifications resulted in a virtually complete loss of HDP activity as measured by: (a) isothermal unwinding of poly [d(A-T)·d(A-T)]; (b) lowering of the melting temperature of DNA; and (c) DNA-binding activity. Figure 4 shows that nitration of less than 2 tyrosines per subunit resulted in loss of melting activity; there was also a concomitant loss of DNA-binding activity (data not shown). Furthermore, nitration of HDP under similar conditions but in the presence of either single stranded DNA or poly d(T) failed to modify the protein, as measured by spectrophotometry and amino acid analysis, and almost complete retention of activity was observed. These data show the involvement of tyrosine(s) in the activity of HDP.

To identify the tyrosine(s) critical in the activity of HDP, equal aliquots of the protein were subjected to nitration and radioiodination (^{125}I) to a level sufficient to inactivate it. Unmodified and modified HDP aliquots then were subjected to tryptic digestion and fractionation by high-performance liquid chromatography (HPLC). Figure 5 compares the three resultant chromatographic profiles. The data show that the yield of a peptide designated as T5 (dashed vertical line) is significantly diminished in the digests of the modified aliquots. Further, new nitrated and radioiodinated peaks appeared in the appropriate profiles at the expense of the T5 peak. In the iodinated sample, the two new peaks contained about 80% of the total radioactivity. Amino acid composition and sequence analysis of the control T5 peak and [^{125}I]-T5 peak showed them to be pure and identical in sequence:

(NH_2)-Glu-Val-Val-Asp-Ser-Ala-Tyr-Glu-Val-Ile-Lys-(COOH)

The other [^{125}I]T-5, 14' peak contained two additional amino acids, Leu and Lys, in addition to those of the T5 peptide. Thus, we identify the T5 peptide as representing the active-site of HDP involved in DNA binding and destabilizing activity.

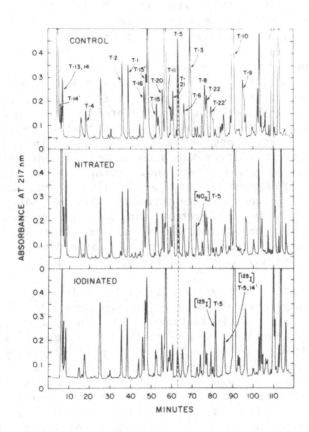

Figure 5. High performance liquid chromatography of HDP pepties. Equal aliquots of control HDP and HDP nitrated and iodinated to a level sufficient to inactive it were digested for 24 hr with trypsin in 8 M urea and 0.1 M NH$_4$CO$_3$. Digests were diluted 4X with water and injected directly onto a Waters C-18 μBondapak column equilibrated with 10 mM potassium phosphate, pH 2.5. Peptides were eluted with increasing concentrations of acetonitrile as follows: 0-30% (86 min.); 30-60% (43 min.) and 60-100% (14 min.).

Structural and Functional Similarities of Rat HDP to Lactate Dehydrogenase-5

With the putative amino acid sequence of the DNA-binding site of rat HDP in hand we wondered whether similar sequences might exist in other DNA-binding proteins. A request to the National Biomedical Research Foundation (Washington, D.C.) to search their computerized Protein Data Base revealed, to our great surprise, that the NADH binding site of LDH-5 (muscle/liver specific M_4 isozyme) had an identical amino acid sequence. Further comparison of sequence data on additional HDP peptides, which comprise about 35% of the protein, with the published sequences of porcine LDH subunits (10) showed the HDP to be surprisingly similar to the M subunit.

Prompted by this apparent sequence homology, we compared rat HDP to various LDHs (rabbit LDH-5, bovine LDH-1 and chicken LDH procured from Sigma Chemical Co. and rat liver LDH-5 purified in this laboratory as described below) with respect to enzymic, DNA-binding and helix-destabilizing activities, and immunological cross-reactivity. Our standard preparations of HDP catalyze the same reduction of pyruvate with NADH as coenzyme as the authentic LDHs do. Using a reaction mixture of 0.1 M tris, pH 8.0, 1 mM pyruvate and 0.15 mM NADH (11), we found that HDP had a specific activity of 220 μ/mg, compared to a specific activity of 180 μ/mg for rabbit LDH-M (Sigma).

To provide the closest possible comparisons of HDP with LDH, it was necessary to purify rat liver LDH-5. We used a modification of the method of Lee et al. (11). About 25 g of rat liver were homogenized in a Sorvall Omnimixer in 75 ml of 10 mM potassium phosphate, pH 6.5, containing 5 mM mercaptoethanol. After centrifugation at 30,000 x g for 1 hr at 4°, the supernatant was applied to a 4 ml column of N^6-(6-aminohexyl) AMP Sepharose (Sigma). Unbound proteins were removed by washing with buffer; the LDH activity was then eluted with buffer containing reduced NAD^+-pyruvate adduct (A_{340}=2). The original procedure (11) included subsequent ion exchange chromatography on Whatman DE-52 to separate LDH-5 from other

LDH isozymes, but we omitted this step in later preparations because we found by staining starch gels for LDH
activity that the AMP-Sepharose eluate contained only
LDH-5. However, the AMP Sepharose eluate contained other
contaminating proteins which gave bands in SDS-PAGE of
40,000 and 50,000 daltons in addition to the band which
had the same mobility as HDP and rabbit LDH-M standards.
We removed these contaminating proteins from LDH preparations by chromatography on single stranded DNA cellulose
as used in HDP purification. Only LDH, and not the
contaminants, was retained on the column; it was eluted
from DNA-cellulose with Buffer A containing 2 M NaCl.

We compared our preparation of rat liver LDH-5 and
the commercially obtained LDHs with HDP in the isothermal
denaturation assay of mixing equal volumes of poly
[d(A-T)· d(A-T)] and protein solutions. Only HDP, rat
liver LDH-5 and rabbit LDH-5 produced significant and
similar increases in absorbance when mixed with a poly
[d(A-T)·d(A-T)] solution. No hyperchromicity was evident with either bovine LDH-1 or the chicken liver LDH
preparation. In addition, rat liver LDH-5 and HDP
gave identical Tm depression values for poly [d(A-T)·
d(A-T)] in Buffer B containing NaCl at concentrations of
20 to 50 mM (data not shown).

An alternative assay for DNA-protein interactions is
to separate DNA-protein complexes from unbound DNA by
filtration through nitrocellulose (12). HDP$_2$ and rat
liver LDH-5 were mixed individually with ^{32}P-labeled
denatured DNA at protein/DNA mass ratios of 10 to 50,
allowed to equilibrate, and filtered slowly through
nitrocellulose. At a protein/DNA ratio of 50, both
proteins bound single stranded DNA to a similar extent
but HDP was more effective than LDH in binding at lower
protein/DNA ratios (Fig. 6). If DNA was not denatured,
neither HDP nor LDH bound to it very well. Binding of
LDH and HDP to denatured DNA in this assay was almost
completely inhibited by preincubation of the proteins
with 20 μM NADH (data not shown). Thus, both HDP and rat
liver LDH-5 bind much more efficiently to denatured DNA
than to native DNA and this binding is inhibited by NADH,
the coenzyme of the LDH reaction.

Antisera against rat HDP have been prepared in this
laboratory by standard procedures by immunization of

rabbits. These anti-HDP sera cross-react with rabbit muscle and rat liver LDH-5 when tested by Ouchterlony immunodifffusion test. Thus, in addition to their sequence homology, HDP and LDH-5 appear to be also similar with respect to their activities and immunological cross-reactivity.

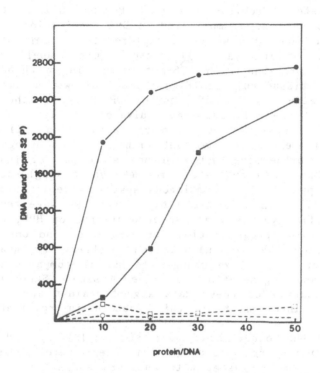

Figure 6. Binding of HDP and LDH-5 purified from rat liver to native and denatured DNA. The reaction mixture (1.0 ml) in Buffer B with 10% glycerol and 10 mM NaCl contained 0.1 µg ^{32}P-labelled native or heat denatured DNA and 0-5 µg HDP or LDH-5. After 30 min. incubation at 25° the mixtures were filtered through nitrocellulose filters (S&S BA85, 0.45 µ) at a flow rate of 1 ml/min. Following two 1 ml washes with the binding buffer, filters were dried and counted to determine the DNA-protein complexes retained on them. Solid lines and filled symbols represent binding to single stranded DNA; dashed lines and open symbols represent binding to double stranded DNA. HDP (• and o); LDH (■ and □).

Function of HDP in vivo

Bacteriophage T4 coded gene 32 protein, which is expressed following infection in E. coli, was the first HDP purified and characterized (15); it remains the most thoroughly studied of all HDPs. Subsequently HDPs have been purified from a variety of prokaryotic and eukaryotic systems (reviewed in 16). Because all of these proteins have been implicated in DNA replication and/or recombination (16), we first explored the pattern of HDP synthesis during the cell cycle. Pulse labelling of cellular proteins with ^{35}S-methionine in synchronized hepatoma tissue culture (HTC) cells, followed by immuno-precipitation of ^{35}S-HDP showed HDP to be synthesized throughout the cell cycle at equivalent levels. Quantitation of total HDP by radioimmunoassay also failed to show differences concommitant with progress through the cycle. Furthermore, HDP did not stimulate or inhibit replication when substitute for the yeast single-strand binding protein in a homologous system of replication in vitro (17). In cell cycle experiments we were testing G1-specific synthesis and/or accumulation of HDP, which may in turn trigger nuclear translocation and onset of replication. We also have failed to observe any quantitative or qualitative changes in HDP in comparisons of normal and regenerating rat liver tissue. Superficial interpretation of these data argued against any easily apparent involvement of this protein in replication.

On the other hand, our study (18) by indirect immunofluorescence staining of polytene chromosomes of Drosophila melanogaster with antisera against rat liver HDP demonstrated concentrations of immunoreactive HDP in many regions of the chromosomes; particularly strong staining was observed in developmental puffs active in transcription. Furthermore, striking concentrations of HDP were detected at specific puffs induced by heat-shock treatment of larvae. The patterns of HDP distribution on these chromosomes were similar to that reported for RNA polymerase (19) and therefore strongly implicated HDP in a transcriptional role. To rule out that staining of puffs was not merely a consequence of facilitated access of antibody to pre-existing HDP and to ascertain additionally whether HDP also might accumulate at replication sites, we have examined Sciara caprophila salivary gland chromosomes by the immunofluorescent staining procedure. At certain developmental stages Sciara chromosomes exhibit distinct replication puffs. As identified by ^{3}H-thymidine labelling, these replication puffs showed little HDP immunofluorescence, while other transcriptional puffs in the same chromosomes exhibited intense staining.

In the aggregate these data suggest that the HDP which we have isolated may be involved in some chromatin-DNA unwinding or melting function related to transcription. However, our subsequent discovery of structural and functional homologies between HDP and LDH now raises several intriguing possibilities. These include: (a) LDH under certain physiological conditions may also function in regulation as a helix-destabilizing protein; (b) the puff-associated immunoreactive material on polytene chromosomes may represent a distinct protein which shares antigenic determinants with LDH or may represent material stained by antibodies against unde-tected minor contaminants in the HDP preparation used as an antigen; or (c) the in vitro activities of LDH/HDP may be artifacts of the assay. Clearly, the available data do not yet allow us to distinguish among these, or other, alternatives.

Conclusion

Our comparison of HDP to LDH is not complete. Yet data presented here leave little doubt that the two activities reside in the same protein molecule. Although we do not know whether the DNA-binding property of LDH-5 is physiologically significant, evidence indicates that DNA-binding proteins studied by other investigators may also be LDH. A Xenopus laevis ovarian protein, which comprises two percent of soluble protein, binds single stranded DNA and stimulates DNA polymerase α_2 reaction in vitro also behaves as LDH (personal communication; Robert Benbow, Johns Hopkins University). The calf thymus protein (33,000 daltons) reported by Herrick and Alberts (4) has amino acid composition resembling LDH. We suspect that single DNA-binding proteins isolated from mouse cells (20, 21) are also LDH. If confirmed, these data suggest that certain cellular proteins may have more than one function. In this context it is noteworthy that epidermal growth factor receptor has been reported recently (22) to have partial DNA-topoisomerase II activity. Clearly, the idea of multifunctional proteins must await additional evidence to become a biological reality.

Acknowledgment

This work was supported by grants from NIH (CA31138) and NSF (PCM-8310692) to GLP and from NIH (GM31539) to KRW.

References

1. Geider, K. and H. Hoffman-Berling. 1981. Proteins Controlling the Helical Structure of DNA. Ann. Rev. Biochem. 50:233-260.

2. Weisbrod, S. 1982. Active Chromatin. Nature 297:289-300.

3. Von Hippel, P.H. and J.D. McGhee. 1972. DNA-Protein Interactions. Ann. Rev. Biochem. 41:231-300.

4. Herrick, G. and B. Alberts. 1976. Purification and Physical Characterization of Nucleic Acid Helix-unwinding Proteins from Calf Thymus. J. Biol. Chem. 251:2124-2132.

5. Alberts, B. and G. Herrick. 1971. DNA-cellulose Chromatography. Meth. Enzymol. 210:198-217.

6. Héléne, C. and G. Lancelot. 1982. Interactions between Functional Groups in Protein-Nucleic Acid Associations. Prog. Biophys. & Mol. Biol. 39:1-68.

7. Reddigari, S.R. 1983. Involvement of Tyrosyl Residues in the Activity of a Helix Destabilizing Protein from Rat Liver. Ph.D. Dissertation. University of Georgia.

8. Riordan, J.F. and B.L. Vallee. 1972. Nitration with Tetranitromethane. Meth. Enzymol. 25:515-521.

9. Fanning, T.G. 1975. Iodination of Escherichia coli lac Repressor. Effect of Tyrosine Modification on Repressor Activity. Biochemistry 14:2512-2520.

10. Kiltz, H., W. Keil, M. Griesback, K. Petry, and H. Meyer. 1977. The Primary Structure of Porcine Lactate Dehydrogenase: Isozymes M_4 and H_4. Hoppe-Seyler Z. Physiol. Chim. 358:123-127.

11. Lee, C-Y., J.H. Yuan, and E. Goldberg. 1982. Lactate Dehydrogenase Isozymes from Mouse. Meth. Enzymol. 89:351-358.

12. Riggs, A.D., H. Suzuki, and S. Bourgeois. 1970. lac Repressor-Operator Interaction. J. Mol. Biol. 48:67-83.

13. Siegel, L.M., and K.J. Monty. 1966. Determination of Molecular Weights and Fractional Ratios of Proteins in Impure Systems by Use of Gel Filtration and Density Gradient Centrifugation. Application to Crude Preparations of Sulfite and Hydroxylamine Reductases. Biochim. Biophys. Acta 112:346-362.

14. Babul, J. and E. Stellwagen. 1969. Measurement of Protein Concentration with Interferences Optics. Anal. Biochem. 28:216-221.

15. Alberts, B.M. and L. Frey. 1970. T4 Bacteriophage Gene 32: A Structural Protein in the Replication and Recombination of DNA. Nature 227:1313-1318.

16. Coleman, J.E. and J.L. Oakley. 1980. Physical Chemical Studies of the Structure and Function of DNA Binding (Helix-Destabilizing) Proteins. CRC Critical Rev. Biochem. 7:247-289.

17. Kojo, H., B.D. Greenberg, and A. Sugino. 1981. Yeast 2-μm plasmid DNA replication in vitro: Origin and Direction. Proc. Natl. Acad. Sci. 78:7261-7265.

18. Patel, G.L. and P.E. Thompson. 1980. Immuno-
     reactive helix-destabilizing protein localized in
     transcriptionally active regions of Drosophila
     polytene chromosomes. Proc. Natl. Acad. Sci. USA
     77:7649-6753.

19. Jamrich, M., A.L. Greenleaf, and E.K.F. Bautz.
     1977. Localization of RNA polymerase in polytene
     chromosomes of Drosophila melanogaster. Proc.
     Natl. Acad. Sci. 74:2079-2083.

20. Otto, B., M. Baynes, and R. Knippers. 1977. A
     Single-Strand-Specific DNA-Binding Protein from
     Mouse Cells that Stimulates DNA Polymerase. Its
     Modification by Phosphorylation. Eur. J. Biochem.
     73:17-24.

21. Seifart, K.H., P.P. Juhasz, and B.J. Benecke.
     1973. A Protein Factor from Rat-Liver Tissue
     Enhancing the Transcription of Native Templates by
     Homologous RNA Polymerase B. J. Biochem. 33:
     181-191.

22. Mroczkowski, B., G. Mosig and S. Cohen. 1984. ATP-
     stimulated interaction between epidermal growth
     factor receptor and supercoiled DNA. Nature 309:
     270-273.

ROLE OF NONHISTONE PROTEIN PHOSPHORYLATION

IN THE REGULATION OF MITOSIS IN MAMMALIAN CELLS

Ramesh C. Adlakha[1], Chintaman G. Sahasrabuddhe[2],

David A. Wright[3], Hélène Bigo[3] and Potu N.Rao[1]

Departments of [1]Chemotherapy Research,

[2]Pathology and [3]Genetics. University of Texas

M. D. Anderson Hospital and Tumor Institute,

Houston, Texas 77030.

The postsynthetic modification of proteins via reversible phosphorylation-dephosphorylation by phosphoprotein kinases and phosphatases has been reported to be an important mechanism in the regulation of numerous intracellular events (15,26,30). A strong correlation between the phosphorylation of histone H1 and (H3) and chromosome condensation and initiation of mitosis has been shown by several investigators (7,8,11,13,14,18,21,23). More recently, it was shown that histone H1 is also phosphorylated during premature chromosome condensation (8,22,27). However, many investigators believe that the superphosphorylation of histone H1 alone is not sufficient for chromosome condensation (10,20,22,27,38). Furthermore, chromosome decondensation could still occur even when dephosphorylation of histone H1 was blocked in mitotic cells (38). There is increasing evidence to suggest that phosphorylation of nonhistone proteins (NHP) may also be required for mitosis-related events. For example, phosphorylation of high mobility group (HMG) proteins has been suggested to be responsible for the shutting off of gene transcription during mitosis (12,28,34), increased phosphorylation of intermediate filament proteins like vimentin at mitosis

59

(17,33), phosphorylation and dephosphorylation of laminar proteins have been implicated in the dissolution and re-formation of the nuclear envelope (19,25). Phosphoryla-tion of ribosomal protein S6, and increased phosphorylation of some other nonhistone proteins including maturation promoting factor have been shown to be associated with meiotic maturation of Xenopus laevis oocytes (29,31,41). Our own studies have suggested a role for the phosphoryla-tion-dephosphorylation reaction mechanism during the maturation of oocytes induced by mitotic factors from HeLa cells (1-4). Recently, we showed that phosphorylation of NHP extractable with 0.2 M NaCl is causally related with the entry of cells into mitosis on one hand, and their dephosphorylation with the exit of cells from mitosis on the other (1,35). Furthermore, the monoclonal antibodies raised against mitotic cells in our laboratory (16) re-cognize a family of phosphoproteins in mitotic cells. The antigenicity of these proteins is destroyed if they are treated with alkaline phosphatase. Thus, the antibodies react with these proteins only when they are phosphoryla-ted. These proteins were not recognized in G1 and S phase cells, probably because they were not phosphorylated. In this study we present additional evidence for the involve-ment of phosphorylation-dephosphorylation of these NHP in the regulation of mitosis in HeLa cells.

CELL CYCLE-SPECIFIC CHANGES IN PHOSPHORYLATION OF NHP

Since the maturation-promoting activity (MPA) of the mitotic factors (proteins extractable with 0.2 M NaCl from mitotic HeLa cells, that induce germinal vesicle breakdown and chromosome condensation when injected into Xenopus laevis oocytes (37)) is greatly stabilized by the presence of phosphatase inhibitors (3,4), we recently carried out a detailed study on the phosphorylation of NHP during the HeLa cell cycle (35). HeLa cells synchronized in S phase by double thymidine block were labeled with ^{32}P at the end of S phase and incubation continued, and the cells sub-sequently collected while they were in G2, mitosis, G1, or S. Cytoplasmic, nuclear, or chromosomal proteins were extracted and the radioactivity incorporated into NHP measured. The phosphorylation levels and the rates of phosphorylation of both the cytoplasmic and chromatin-binding NHP increased slowly during early- to mid- G2 but rapidly during late-G2 and reached a peak in mitosis. An

8-10 fold increase was observed in the phosphorylation of
NHP from mid- G2 to mitosis. When the cells divided and
entered G1, the amount of radioactivity in the NHP of G1
cell extracts dramatically decreased to about 10% of that
of the mitotic cell extracts. The presence of cycloheximide during the reversal of mitotic block had no effect
on either the rate of exit from mitosis to G1 or the loss
of radioactivity from the NHP. The rate of phosphorylation
of NHP was extremely low during G1 and decreased further
(to 3-5% of that in mitotic cells) in S phase; it again
increased gradually during early G2, reaching a peak in
mitosis (35). Westwood et al. (40) reported recently
results very similar to ours on the phosphorylation of
NHP during the cell cycle of CHO cells. However, Song and
Adolph (36), also working with HeLa cells, reported that
NHP from isolated metaphase chromosomes are strikingly dephosphorylated in comparison to those of S phase chromatin.
The differences between these two studies most likely stem
from the different experimental protocols used for the
extraction of NHP, and the two studies therefore, investigated phosphorylation in altogether different subsets of
NHP.

IDENTIFICATION OF THE NHP PHOSPHORYLATED

Proteins from the extracts of mid-G2, mitotic, and
early-G1 cells labeled with ^{32}P were separated by SDS-
polyacrylamide gel electrophoresis. Autoradiography of the
gel showed that eight major proteins (with apparent molecular masses of 100,92,70,55,43,36 and 27.5 kd) were extensively phosphorylated at mitosis. The relative amount
of ^{32}P incorporated into NHP in mid G2 and early- G1 cells
was only 10-20% of that in mitotic cells (35). These
studies suggest that phosphorylation of NHP may play a role
in mitosis.

EFFECT OF X-IRRADIATION, AND CYCLOHEXIMIDE ON G2-TRAVERSE
AND PHOSPHORYLATION OF NHP

To determine whether X-ray-induced mitotic delay would
be associated with delay in NHP phosphorylation, HeLa cells
synchronized in G2 phase were exposed to 500 rad of X-rays
and incubation continued in the presence of Colcemid and
^{32}P. A 3.5 h delay in accumulated mitotic index induced
by X-irradiation resulted in a corresponding delay in the

incorporation of ^{32}P into NHP (35).

Synthesis of new proteins has long been known to be
necessary for the G2-M transition (39). Our own studies
have shown that mitotic factors become available only
during G2 and that they reach a critical level by the G2-
M transition. Therefore, we decided to determine if these
newly synthesized proteins were phosphorylated at the G2-
M transition. HeLa cells synchronized in S phase were
labeled with ^{32}P beginning at the end of S phase and
aliquots of cells were pulse treated with cycloheximide for
1 h at different times during G2. While the level of
phosphorylation of NHP increased with time in the control
cells, it ceased to increase in the treated cells as soon
as cycloheximide was added (35). Why should cycloheximide
block phosphorylation? If cycloheximide were blocking
synthesis of enzymes necessary for phosphorylation, i.e.,
phosphokinases, the enzymes existing prior to the addition
of cycloheximide should be active and continue to phosphorylate NHP, unless they have an extremely short half-
life. Alternatively, NHP phosphorylation observed during
G2 occurs in the newly synthesized NHP. Therefore, it
appears that continued synthesis and immediate phosphorylation of NHP may be equally important for the G2-M
transition.

LACK OF NHP PHOSPHORYLATION IN G2-ARRESTED CELLS

Al-Bader et al. (9) have previously shown that cells
arrested in G2 by treatment with cis-acid (cis-4[[[(2-
chloroethyl)-nitrosoamino]carbonyl]amino]cyclohexane car-
boxylic acid) lack certain G2-specific proteins. We have,
therefore, examined the phosphorylation of NHP in cis-acid
treated cells. HeLa cells synchronized in S phase were
treated with cis-acid for 1 h as described earlier (9,35).
After washing cells free from cis-acid, they were incubated
in fresh medium containing Colcemid. After a 20 h incubation, the rounded and loosely attached mitotic cells were
removed by selective detachment and both the G2-arrested
and mitotic populations were separately incubated with ^{32}P.
Cell samples were taken at 2 h intervals, NHP extracted,
and incorporation of label into NHP determined. The amount
of label incorporated into NHP over a period of 8 h in the
mitotic cells was four-fold greater than in the G2-arrested
cells (35). These data further support our conclusion

that NHP phosphorylation is closely associated with the entry of cells into mitosis.

ASSOCIATION OF UV-INDUCED CHROMOSOME DECONDENSATION WITH DEPHOSPHORYLATION OF NHP IN MITOTIC HeLa CELLS

UV-irradiation of mitotic HeLa cells, if followed by an incubation, has been known to result in the attenuation or decondensation of metaphase chromosomes. Furthermore, presence of hydroxyurea and Ara C during the postirradiation incubation greatly enhances the UV-induced decondensation of metaphase chromosomes (6,24). We have also shown that extracts obtained from UV-treated mitotic cells, incubated for 2 h following irradiation with 2,500 erg of UV in the presence of hydroxyurea and Ara-C, had very little MPA. Furthermore, the inactivation of mitotic factors (or loss of MPA) was shown to be specifically associated with the UV-induced chromosome decondensation. Very little inactivation of mitotic factors was observed even with very high doses of X-irradiation (6). We, therefore, examined the relationship between UV-induced chromosome decondensation and dephosphorylation of NHP. HeLa cells synchronized in mitosis by N_2O block method (32) were divided into two groups; one group was irradiated with 2,500 erg of UV and incubated in fresh medium containing hydroxyurea ($10^{-2}M$), Ara-C ($10^{-4}M$), Colcemid and ^{32}P and the other group was kept as control and incubated with ^{32}P and Colcemid only. Cell samples were taken at different times and processed for incorporation of label into NHP. The rate of phosphorylation of NHP was significantly lower in UV-treated cells as compared with the control and decreased further with time (Table 1). Accordingly, the rate of dephosphorylation of NHP in UV-irradiated mitotic HeLa cells prelabeled with ^{32}P during prolonged mitotic arrest was significantly higher than in control mitotic cells (Table 2). NHP that were dephosphorylated during the course of UV-induced chromosome decondensation were identified by SDS-polyacrylamide gel electrophoresis followed by autoradiography (Fig. 1). A much clearer picture of the NHP that get dephosphorylated, was obtained when mitotic factor activity from the ^{32}P labeled control and UV-treated mitotic cells was purified approximately 200-fold by affinity chromatography on DNA-cellulose (Adlakha, Wright, Sahasrabuddhe and Rao, manuscript in preparation) and separated by gel electrophoresis and

TABLE 1

LEVEL OF PHOSPHORYLATION OF CONTROL AND UV-IRRADIATED
MITOTIC HeLa CELLS DURING PROLONGED MITOTIC ARREST

Incorporation of ^{32}P into NHP (cpm/10^6cells)

Hours held in mitosis	NHP in Cytoplasmic Fraction			NHP in Chromosomal Fraction		
	Control	UV-treated	% of control	Control	UV-treated	% of control
1	739	589	79.7	580	359	61.9
2	1240	696	56.1	1110	505	45.5
3	3090	1451	46.9	2322	1106	47.6
4	6282	3044	48.4	4777	2533	53.0

TABLE 2

RATE OF DEPHOSPHORYLATION OF NHP IN UV-IRRADIATED MITOTIC
HeLa CELLS PRELABELED WITH ^{32}P DURING PROLONGED MITOTIC
ARREST

Incorporation of ^{32}P into NHP (cpm/10^6 cells)

Hours held in mitosis	NHP in Cytoplasmic Fraction			NHP in Chromosomal Fraction		
	Control	UV-treated	Relative loss (%)	Control	UV-treated	Relative loss (%)
0	189,083	189,083	0	163,493	163,493	0
1	175,750	141,200	19.6	150,533	137,450	8.7
2	157,614	100,160	36.45	143,054	101,854	28.8
4	135,187	72,187	46.60	127,183	75,038	41.0

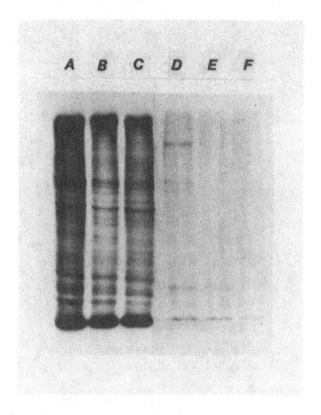

Fig. 1. Relationship between UV induced chromosome decondensation and dephosphorylation of NHP. ^{32}P labeled NHP
from control and UV-irradiated mitotic HeLa cells were subjected to electrophoresis on SDS-polyacrylamide gels and
gels autoradiographed. Lanes: A, mitotic extract (control);
B, extract from UV-irradiated mitotic cells (30 min incubation); C, (2 h incubation; D and E, mixture of ^{32}P labeled
mitotic and unlabeled G1 or S phase cell extracts, respectively. M. Wt. St; Molecular weight standards. Note the
decrease in intensity of labeling in lanes B, C and D.

autoradiography (compare lanes D and E in Fig. 1). These
data suggest that UV-irradiation of mitotic HeLa cells results in the induction or activation of phosphatases that
may specifically dephosphorylate mitotic NHP. A similar
dephosphorylation of these mitotic NHP was also observed
when partially purified prelabeled mitotic factors were

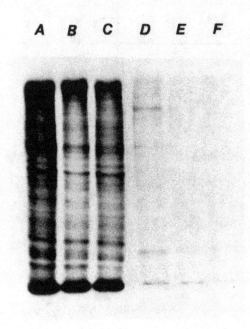

Fig. 2. ^{32}P labeled NHP from control and UV-irradiated
mitotic HeLa cells were enriched ($\sim$200-fold) for mitotic
factor activity, by affinity chromatography on DNA-cellulose, separated by electrophoresis on SDS-polyacrylamide
gels and gels autoradiographed. Lanes: A, B and C same as
lanes A, C and D, respectively in figure 1; D,enriched
fraction from ^{32}P labeled mitotic extract; E, corresponding
fraction from extracts of UV-irradiated mitotic cells; F,
mixture of ^{32}P labeled enriched fraction (lane D) and unlabeled G1 cell extract.

mixed with G1 cell extracts, but not with S phase cell extracts (lane F in Fig. 2), indicating the presence of
similar, if not, identical phosphatases in G1 cells. We
have previously shown that mixing of mitotic cell extracts
with G1 cell extracts results in the inactivation of mitotic

factors under the influence of the inhibitory factors present in G1 cells (5). Furthermore, inhibitors of mitotic factors (IMF) can be induced both in G_0 phase and mitotic cells by UV irradiation (5,6). Taken together, these results strongly suggest that phosphorylation-dephosphorylation of this specific subset of NHP may represent the molecular mechanism for the regulation of chromosome condensation and decondensation in the life cycle of eukaryotic cells.

ACKNOWLEDGEMENTS

We thank Josephine Neicheril for her excellent secretarial assistance in the preparation of this manuscript. This study was supported in part by research grants CA-27544, CA-21927 from NCI, CH-205 from the American Cancer Society, BR-86 from this Institute and a grant from Kleberg Foundation.

REFERENCES

1. Adlakha, R.C., F.M. Davis, and P.N. Rao. 1984. In Cell Proliferation: Recent Advances, Vol. 1. A.L. Boynton and H.L. Leffert, Editors. Academic Press, Inc., New York. In press.
2. Adlakha, R.C., C.G. Sahasrabuddhe, and P.N. Rao. 1982. J. Cell Biol. 95:75a.
3. Adlakha, R.C., C.G. Sahasrabuddhe, D.A. Wright, W.F. Lindsey, and P.N. Rao. 1982. J. Cell Sci. 54:193-206.
4. Adlakha, R.C., C.G. Sahasrabuddhe, D.A. Wright, W.F. Lindsey, M.L. Smith,and P.N. Rao. 1982. Nucleic Acids Res. 10:4107-4117.
5. Adlakha, R.C., C.G. Sahasrabuddhe, D.A. Wright, and P.N. Rao. 1983. J. Cell Biol 97:1707-1713.
6. Adlakha, R.C., Y.C. Wang, D.A. Wright, C.G. Sahasrabuddhe, H. Bigo, and P.N. Rao. 1984. J. Cell Sci. 65:279-295.
7. Ajiro, K., T.W. Borun, and L.H. Cohen. 1981. Biochemistry. 20:1445-1454.
8. Ajiro, K., T. Nishimoto, and T. Takahashi. 1983. J. Biol. Chem 258:4534-4538.
9. Al-Bader, A.A., A. Orengo, and P.N. Rao. 1978. Proc. Natl. Acad. Sci. USA 75:6064-6068.
10. Allis, C.D., and M.A. Gorovsky. 1981. Biochemistry. 20:3828-3833.
11. Balhorn, R., V. Jackson, D. Granner, and R. Chalkley.

1975. Biochemistry. 14:2504–2511.

12. Bhorjee, J.S. 1981. Proc. Natl. Acad. Sci. USA 78: 6944–6948.

13. Bradbury, E.M., R.J. Inglis, H.R. Matthews,and T.A. Langan. 1974. Nature (London) 249:553–556.

14. Bradbury, E.M., R.J. Inglis, H.R. Matthews, and N. Sarner. 1973. Eur. J. Biochem. 33:131–139.

15. Cohen, P. 1982. Nature (London) 296:613–620.

16. Davis, F.M., T.Y. Tsao, S.K. Fowler, and P.N. Rao. 1983. Proc. Natl. Acad. Sci. USA 80:2926–2930.

17. Evans, R.M., and L.M. Fink. 1982. Cell. 29:43–52.

18. Fisher, S.G., and U.K. Laemmli. 1980. Biochemistry. 19:2240–2246.

19. Gerace, L., and G. Blobel. 1980. Cell. 19:277–287.

20. Gorovsky, M.A., and J.B. Keevert. 1975. Proc. Natl. Acad. Sci. USA 72:2672–2676.

21. Gurley, L.R., R.A., Tobey, R.A. Walters, C.E. Hildebrand, P.G. Hohmann, J.A. D'Anna, S.S. Barham, and L.L. Deaven. 1978. In Cell Cycle Regulation, J.R. Jeter, I.L. Cameron, G.M. Padilla, and A.M. Zimmerman, Editors, Academic Press, Inc., New York. 37–60.

22. Hanks, S.K., L.V. Rodriguez, and P.N. Rao. 1983. Expt. Cell Res. 148:293–302.

23. Johnson, E.M., and V.G. Allfrey. 1978. In Biochemical Action of Hormones, G. Litmack, Editor, Vol. 5 Academic Press, Inc., New York. 1–51.

24. Johnson, R.T., A.R.S. Collins, and C.A. Waldren. 1982. In Premature Chromosome Condensation: Applications in Basic, Clinical, and Mutation Research, P.N. Rao, R.T. Johnson, and K. Sperling, Editors. Academic Press, Inc., New York. 253–308.

25. Jost, E., and R.T. Johnson. 1981. J. Cell Sci. 47:25–53.

26. Krebs, E.G., and J.A. Beavo. 1979. Annu. Rev. Biochem. 48:923–959.

27. Krystal, G.W., and D.L. Poccia. 1981. Expt. Cell Res. 134:41–48.

28. Levy-Wilson, B. 1981. Proc. Natl. Acad. Sci. USA 78:2189–2193.

29. Maller, J.L., M. Wu, and J.C. Gerhart. 1977. Dev. Biol. 58:295–312.

30. Nestler, E.J., and P. Greengard. 1983. Nature (London) 305:583–588.

31. Nielsen, P.J., G. Thomas, and J.L. Maller. 1982. Proc. Natl. Acad. Sci. USA 79:2937–2941.

32. Rao, P.N. 1968. Science (Wash. D.C.). 160:774-776.
33. Robinson, S.I., B. Nelkin, S. Kaufmann, and B. Volgelstein. 1981. Expt. Cell Res. 133:445-449.
34. Safer, J.D., and R.I. Glazer. 1982. J. Biol. Chem. 257:4655-4660.
35. Sahasrabuddhe, C.G., R.C. Adlakha, and P.N. Rao. 1984. Expt. Cell Res. In press.
36. Song, M.K.H., and K.W. Adolph. 1983. J. Biol. Chem. 258:3309-3318.
37. Sunkara, P.S., D.A. Wright, and P.N. Rao. 1979. Proc. Natl. Acad. Sci. USA 76:2799-2802.
38. Tanphaichitr, N., K.C. Moore, D.K. Granner, and R. Chalkley. 1976. J. Cell Biol. 69:43-50.
39. Tobey, R.A., D.F. Peterson, E.C. Anderson, and T.T. Puck. 1966. Biophys. J. 6:567-581.
40. Westwood, J.T., E.B. Wagenaar, and R.B. Church. 1983. J. Cell. Biol. 97:147a.
41. Wu, M., and J. C. Gerhart. 1980. Dev. Biol. 79: 465-477.

NONHISTONE PROTEINS, FREE RADICAL DEFENSES AND

ACCELERATION OF SPHERULATION IN PHYSARUM

C. Nations, R.G. Allen and J. McCarthy

S.M.U. Biology Department

Dallas, Texas 75275

INTRODUCTION

The syncytial slime mold Physarum polycephalum has been widely employed as a model for the study of a variety of cellular events; however, the overwhelming emphasis in modern Physarum research has been on those areas directly related to growth and differentiation (22). Mitosis is synchronous following an interphase period of 8-10 h and mitotic synchrony implies metabolic synchrony (13). All events involved in the regulation of differentiation are therefore presumed to be amplified since they should occur simultaneously. This hypothesis has been the basis for intensive efforts to identify the role of specific enzymes (15), nucleic acids (2,8), and nuclear proteins (17) in the developmental processes of eucaryotes. A considerable volume of literature concerning developmental processes in Physarum has accumulated since Rusch (22) first introduced the organism as a model system for probing the events that govern the life cycle of cells. At this time little conclusive information has resulted; endogenous events and factors which induce specific cell state transitions have not been identified. A perennial problem has been to distinguish between the events that are causative and the events which are the consequence of differentiation. For example, when Physarum differentiates into groups of dormant spherules (sclerotia) dramatic changes occur in the electrophoretic profile of the nonhistone chromosomal proteins; however, it has not been possible to determine

71

whether these changes induce differentiation or are a
parallel response to the inductive treatment (17). The
recent discovery of a white mutant strain of Physarum (LU
887xLU897) may provide a key to the solution of this
problem (1). Our preliminary studies indicate that subjecting the white organism to a routine starvation treatment
(7) frequently fails to induce the formation of dormant
material. Microscopic examination of the white microplasmodia has revealed that the organism does not
differentiate into microsclerotia during starvation. Thus,
the white strain may provide a much needed experimental
control for investigation of the events which result in
differentiation.

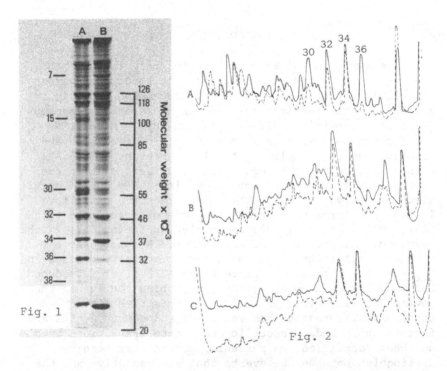

Fig. 1.   Nonhistone proteins of (A) yellow and (B) white
          Physarum microplasmodia.
Fig. 2.   Nonhistones of yellow (—) and white (--) micro-
          plasmodia after (A) two; (B) three and (C) four
          days of growth (17).

SPHERULATION AND NONHISTONE PROTEINS

Spherulation is a form of diploid encystment. *Physarum* microplasmodia grown in shake flasks cleave into multinucleated spherules when treated with mannitol or transferred to a salts-only starvation medium. When mannitol is used to induce spherulation, the spherules become separate units; starvation-induced spherules remain as clustered microsclerotia (5,7).

The hypothesis that nonhistone proteins regulate gene expression has gained considerable popularity. Since the yellow (M3cVII) strain of *Physarum* forms microsclerotia as it depletes the nutrient medium and the white strain does not, it was of interest to compare the nonhistone proteins of the two strains. As seen in figure 1, the electrophoretic spectra of nonhistone proteins are quite similar in the two strains during logarithmic growth. The starvation-induced changes that have been suggested to trigger differentiation in the yellow strain are already evident in the white strain during exponential growth (20). In the yellow strain the polypeptides represented by bands 30 and 36 decrease sharply with the onset of differentiation; these bands are almost nonexistant in the white strain under all conditions of growth. Nuclear actin (band 32) has been suggested to be the most important nonhistone protein involved in a general chromatin inactivation mechanism that leads to a quiescent cell state (17); this protein is also prominent in the white strain throughout the vegetative phase (Fig. 2A-C).

OXYGEN METABOLISM, FREE RADICALS AND DIFFERENTIATION

Cellular differentiation ultimately results from the differential expression of genes; however, this process is not independent of cytosolic influence. Gurdon et al. (12) showed that transplanting frog somatic cell nuclei into frog oocytes resulted in the formation of tadpoles. This effect was not due to the dilution of intrinsic gene repressors since transplantation of dedifferentiated nuclei, i.e., nuclei from cancer cells, to oocytes has been shown to result in the formation of normal tissues (19). Although such studies indicate the existence of cytosolic factors which influence gene expression, the nature of cytosolic influence on differentiation is presently obscure. Many

of the changes associated with development appear to
correspond either directly or indirectly with alterations
in oxygen metabolism (11). If oxygen metabolism exerts an
effect on differentiation it would seem improbable that
molecular oxygen is directly involved. Most of the O_2
consumed by aerobic cells is tetravalently reduced to
water; however, a small portion is univalently reduced to
O_2^-, the superoxide free radical (4). Sequential univalent
reductions of oxygen would lead to the production of O_2^-,
hydrogen peroxide (H_2O_2), and the highly reactive hydroxyl
free radical (OH·) (10). Many free radical reactions are
potentially damaging, but cells contain a variety of
defenses which can quench free radicals and thus reduce the
number of damaging reactions. Superoxide dismutase (SOD)
reduces O_2^- to H_2O_2; catalase and peroxidases reduce H_2O_2 to
H_2O. Glutathione peroxidase can eliminate H_2O_2 as well as
the lipid peroxides which are formed during free radical
reactions (4,14). Enzymic defenses which eliminate the OH·
radical are not known; however, the tripeptide glutathione
(GSH) can quench the OH· radical by reacting directly with
it. This reaction results in the oxidation of GSH to GSSG
(10). GSH is also an important factor in determining the
redox state of the cell (4).

Transition metals that catalyse the formation of OH·
radicals induce spherulation in Physarum (16). The free
radical scavenger mannitol is routinely employed to prepare
spherules (5). The respiratory rate of spherulating
microplasmodia declines during the transition (18); this
may suggest that the rate of free radical generation is
altered during differentiation. Child (6) found that the
higher orders of developing organisms contain a metabolic
gradient and postulated that the regions of higher metabolic
activity influence the development of the regions of lower
metabolic activity. Amphibians exhibit a complete shift
from the high oxygen-binding hemoglobin present in tadpoles
to a form of hemoglobin with low oxygen-binding capacity in
adults (11). Caplan and Koutropas (3) demonstrated that
differential vascularization, which would presumably lead
to unequal oxygenation of tissues, occurs in developing
chick embryos. Whether such changes result from differential gene expression or act as a causal factor in the
differentiation of cells is at present unclear. However, it
has been demonstrated that phenotypic expression in
cultured embryonic chick cells can be controlled by growing

the cells under different O_2 tensions (3).

We find that transfer of microplasmodia to the salts medium for 12 h increases the cellular concentration of inorganic peroxides by 80% and causes a 70% decrease in GSH concentration. These changes may in part be accounted for by the observation that H_2O_2 production is greatest during state +4 respiration; H_2O_2 generation is believed to correspond roughly to the rate of free radical generation (9). We have been unable to demonstrate catalase activity in Physarum; however, the organism does exhibit GSH peroxidase activity. Inhibition of GSH synthesis with L-buthionine sulfoximine (BUS) significantly increases the peroxide concentration of microplasmodia in the salts medium; however, this treatment does not increase the level of inorganic peroxides in cultures maintained in growth medium, even though treatment with BUS decreases GSH concentration by 47%. Conversely, a 41% augmentation of GSH concentration, obtained by treatment with L-2-oxothiazolidine-4-carboxylate (LOC), stimulates the production of inorganic peroxides in growth medium but not in the salts medium. The increase in peroxides caused by LOC in growth medium may be due to an increased rate of GSH autoxidation. The failure of LOC to enhance GSH concentration in salts medium is presumably due to GSH oxidation by GSH peroxidase. In most cases cells do not retain GSSG. Also, both LOC and BUS significantly increase the rate of differentiation in Physarum (Fig. 3).

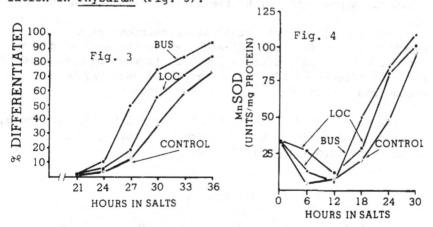

Figs. 3 & 4.   Effects of 4 mM BUS and 2 mM LOC, on differentiation and on mangano-SOD isozyme activity in Physarum.

There are at least two lines of evidence which indicate free radical involvement in tne differentiation process: 1) the salivary glands of several species of insects contain polytene chromosomes which undergo puffing as a normal part of development, and uncouplers of mitochondrial respiration such as dinitrophenol, menadione and antimycin A which are believed to generate free radicals, have been observed to induce chromosomal puffing (9,23); 2) alterations in the differentiated state are invariably accompanied by changes in the level of cellular free radical defenses. Most notably, all cancer cells appear to exhibit a reduction in mitochondrial SOD (mangano-isozyme; 14). In many cases the activity of cytosolic SOD (Cu/Zn isozyme) is also greatly reduced (21). Other free radical quenching enzymes also exhibit decreased activity in dedifferentiated tissue (21).

We have observed a sharp decline in SOD activity in microplasmodia during the first 6-12 h after transfer to salts medium; total SOD activity increases sharply between 6 and 30 h. The change appears to be primarily due to the activity of the mangano-isozyme, which increased 2100% during this period. LOC or BUS-treated cultures exhibit a similar pattern of change in SOD activity except that SOD activity begins to increase at 12 h in these cultures and the rate at which the increase occurs in the treated groups is statistically greater than in the controls (Fig. 4). A comparison of figures 3 and 4 shows that the rate at which the mangano-isozyme of SOD increases in differentiating cultures is proportional to the rate of differentiation.

The results of this study are consistent with the hypothesis that free radical-mediated events play a role in the differentiation process; however, it must be acknowledged that the effects of LOC and BUS reported here may be via mechanisms unrelated to free radical defenses.

We thank Ms. K. Farmer, Mr. P. Toy and Dr. W. Fagerberg, for technical assistance. This work was supported by the Glenn Foundation for Medical Research.

REFERENCES

1. Anderson, R.W. 1977. A plasmodial colour mutation in the myxomycete Physarum polycephalum. Genet. Res. (Camb.) 30, 301-306.

2. Braun, R. 1982. RNA metabolism. In Biology of Physarum
    and Didymium, H.C. Aldrich and J.W. Daniel, editors,
    Academic Press, New York, pp. 393-435.
3. Caplan, A.I. and S. Koutroupas. 1973. The control of
    muscle and cartilage development in the chick limb:
    the role of differential vascularization. J. Embroyl.
    Exp. Morph. 29, 571-583.
4. Chance, B., H. Seis, and A. Boveris.1979. Hydroperoxide
    metabolism in mammalian organs. Physiol. Rev. 59,
    527-603.
5. Chet, I. and H.P. Rusch. 1969. Induction of spherule
    formation in Physarum polycephalum by polyols. J.
    Bacteriol. 100, 674-678.
6. Child, C.M. 1915. Individuation and reproduction in
    organisms. In Scenesence and Rejuvenscence, C.M. Child,
    editor, University of Chicago Press, Chicago, Illinois.
    pp. 199-236.
7. Daniel, J. and H. Baldwin. 1964. Methods of culture for
    plasmodial myxomycetes. In Methods in Cell Physiology,
    D.M. Prescott, editor, Academic Press, New York, 1:9-41.
8. Evans, T.E. 1982. Organization and replication of DNA
    in Physarum polycephalum. In Biology of Physarum and
    Didymium, H.C. Aldrich and J.W. Daniel, editors,
    Academic Press, New York, pp. 371-391.
9. Foreman, H.J., and A. Boveris. 1982. Superoxide radical
    and hydrogen peroxide in mitochondria. In Free Radicals
    in Biology, W.A. Pryor, editor, Academic Press, New
    York, 5:65-90.
10. Forman, J.J., and A.B. Fischer. 1981. Antioxidant
    Defences. In Oxygen and Living Processes, D.L. Gilbert,
    editor, Springer-Verlag, New York, pp. 235-249.
11. Frieden, E. 1981. The dual role of thyroid hormones in
    vertebrate development and calorigenesis. In Meta-
    morphosis, L.I. Gilbert and E. Frieden, editors, Plenum
    Press, New York, N.Y. 2nd edition, pp. 545-563.
12. Gurdon, J.B., R.A. Laskey, and O.R. Reeves. 1975. The
    developmental capacity of nuclei transplanted from
    keratinized skin cells of adult frogs. J. Embroyol.
    Exp. Morph. 34, 93-112.
13. Guttes, E. and S. Cuttes. 1964. Mitotic synchrony in
    the plasmodia of Physarum polycephalum by coalescence of
    microplasmodia. In Methods in Cell Physiology, D.M.
    Prescott, editor, Academic Press, New York, 1:43-54.

14. Halliwell, B. 1981. Oxygen toxicity, free radicals and
    aging. In Age Pigments, R.S. Sohal, editor, Elsevier/
    North Holland, Amsterdam, pp. 1-62.
15. Hutterman, A. 1982. Enzyme and protein synthesis during
    differentiation of Physarum polycephalum. In
    Cell Biology of Physarum and Didymium, H.C. Aldrich and
    J.W. Daniel, editors, Academic Press, New York, pp.
    77-99.
16. Jump, J.A. 1954. Studies on sclerotization in Physarum
    polycephalum. Am. J. Bot. 41, 561-567.
17. Lestourgeon, W.M., R. Totten, and A. Forer. 1974. The
    nuclear acidic proteins in cell proliferation and
    differentiation. In Acidic Proteins of the Nucleus,
    I.L. Cameron and J.R. Jeter, editors, Academic Press,
    New York, pp. 159-190.
18. Lynch, T.J., and Henney, H.R. 1973. Carbohydrate
    metabolism during differentiation (sclerotinization)
    of the myxomycete Physarum polycephalum. Arch.
    Mikrobiol. 90, 189-198.
19. McKinnell, R.G., B.A. Deggins, and D.D. Labat. 1969.
    Transplantation of pluripotential nuclei from triploid
    frog tumors. Science 165, 394-395.
20. Nations, C., and J.L. McCarthy. 1984. Growth of white
    microplasmodia of Physarum polycephalum. Comp. Biochem.
    Physiol. in press.
21. Oberley, L.W. 1983. Superoxide dismutase and cancer.
    In Superoxide Dismutase, L.W. Oberley, editor, CRC
    Press, Boca Raton, Florida, 2:127-165.
22. Rusch, H.P. 1980. Introduction. In Growth and
    Differentiation in Physarum polycephalum, W.H. Dove
    and H.P. Rusch, editor, Princeton University Press,
    Princeton, New Jersey, pp. 1-8.
23. Zegarelli-Schmidt, E.C., and R. Goodman. 1981. The
    dipterian as a model system in cell and molecular
    biology. Internat. Rev. Cytol. 71, 245-363.

COORDINATE AND NON-COORDINATE REGULATION OF

HISTONE mRNAs DURING MYOBLAST GROWTH AND

DIFFERENTIATION

R. Curtis Bird, Fred A. Jacobs and Bruce H. Sells

Department of Molecular Biology and Genetics,

College of Biological Science,
University of Guelph,
Guelph, Ontario, Canada

INTRODUCTION

The fusion of myoblasts to produce non-dividing syncytial myotubes marks the end of cell proliferation during muscle cell differentiation. Following the signal to differentiate myoblasts enter G_1 phase, retire from the cell cycle and differentiate into myotubes (3,4). The synthesis of muscle specific proteins is initiated as the cells begin to differentiate and defines the onset of the differentiated state (2,7). However the conditions necessary and sufficient to allow a proliferating population of cells to withdraw from the cell cycle and to switch from a growth specific to a differentiation specific program of gene expression have yet to be defined.

These studies are designed to define the changes in expression and regulation of genes thought to be cell growth related. Histone gene expression is largely confined to periods of active DNA synthesis (S phase), and cell proliferation and is an example of a gene family which is inactive during periods of non-proliferation (5,9).

79

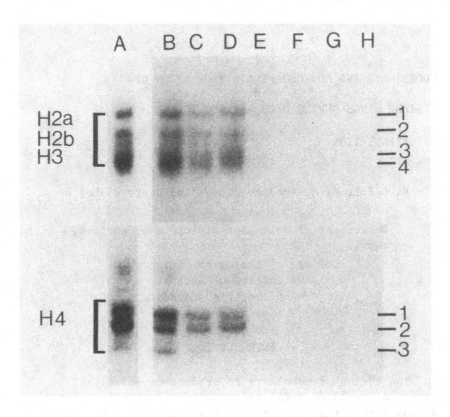

Figure 1.   The abundance of histone mRNA subspecies during
myoblast growth and differentiation.   Polysomal RNA
(50ug/lane) was isolated from cells that were: A)S phase
synchronized, B) 50%, C) 60% or D) 80% confluent, or given
the signal to differentiate for: E) 1, F) 2, G) 3 or H) 4
days.   RNA was separated on a denaturing 6% polyacrylamide
gel, blotted and probed for histone mRNA sequences.

RESULTS

In this investigation the level of various histone
mRNAs was examined during myoblast differentiation, in S
phase synchronized cells and following inhibition of DNA
synthesis with Ara-C or hydroxyurea.   In our initial
experiments histone mRNA content of polysomal RNA from
proliferating L6-5 myoblasts was determined in growing

populations approaching confluency and after myoblasts had been given the signal to differentiate (Fig. 1B-D). Each of the core histone mRNAs was separated into multiple subspecies on denaturing polyacrylamide gels, blotted and probed for the specific histone mRNA sequences. As proliferating populations of L6-5 cells approached confluency the amount of each of the multiple histone mRNA subspecies declined. A tight coordination among all of the various subspecies was observed during the deceleration in cell proliferation.

Proliferating cultures of L6-5 cells were given the signal to differentiate by changing their growth medium to a serum reduced type (10% fetal calf serum to 2.5% horse serum). Under these conditions L6-5 cells have largely retired from the cell cycle by the end of day 1, have begun aligning themselves by the end of day 2 and by day 3 have begun fusing to form terminally differentiated syncytial myotubes. By 48 hr after the signal to differentiate the levels of all of the subspecies of histone mRNA had decreased to less than 5% of exponential growth levels and remain at this basal level during the remainder of the differentiation period. The rate of decline was rapid and coordinate among all histone mRNA subspecies (Fig. 1E-H). Exceptions to this coordinate behaviour were detectable only after further analysis (see below).

To produce an S phase population of cells myoblasts were synchronized by double thymidine block (8). The amount of histone mRNA/ug of polysomes recovered from S phase synchronized cells was greater than that recovered from the polysomes of exponentially growing cells (Fig. 1A). Each subspecies of histone H2a, H2b and H3 mRNA was increased proportionately. In the case of histone H4 the two major mRNA subspecies were nearly 2-fold higher while subspecies 3 remained at approximately the same level. H4 subspecies 3 appeared to be constitutively expressed during S phase and in exponentially growing cells and is not coordinately regulated with the other H4 mRNAs.

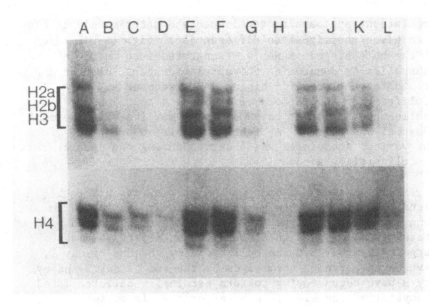

Figure 2. The abundance of histone mRNAs following
inhibition of DNA or RNA synthesis by treatment with:
A-D)Ara-C, E-H)hydroxyurea or I-L)actinomycin D for 0,15,30
or 60 min for each inhibitor respectively. RNAs were
analysed as described (Fig. 1).

    Following myoblast differentiation as described above
a rapid decline occurred in the levels of all of the
histone mRNA subspecies. During the same period the cells
underwent a decline in the rate of DNA synthesis (1). To
determine whether direct inhibition of DNA synthesis in
growing cells affects the level of histone mRNA, myoblasts
were treated with Ara-C or hydroxyurea. Treatment with
either inhibitor resulted in a dramatic decrease in
histone mRNA half-life to 10-13 min in comparison to a
half-life of 38 min following inhibition of RNA transcription with actinomycin D (Fig. 2). Furthermore, there is a
tight coordinate regulation of histone mRNA levels
intimately associated with DNA synthesis. Inhibition of
transcription alone cannot account for the rapid decay rate
observed. A mechanism which actively promotes the
destruction of histone mRNAs seems to be involved since the
histone mRNAs vanish completely from the cytoplasm.

We have identified a single subspecies of H4 mRNA
which is polyadenylated (1). This individual subspecies
was examined during myoblast differentiation and after
inhibition of DNA synthesis to determine whether its
behaviour was similar to the behaviour of the other histone
mRNA subspecies. Following differentiation the poly(A)$^+$
H4 mRNA dropped to 30% of its level in proliferating cells
(Fig. 3). This differs substantially from the greater than
95% drop observed for the other histone mRNA subspecies
(Fig. 1). Nevertheless, the half-life of the poly(A)$^+$
H4 mRNA subspecies (15 min) was similar to that obtained
for the total histone mRNA population following Ara-C
inhibition of DNA synthesis. A measure of non-coordinate
regulation seems to exist among histone H4 mRNA subspecies
during the transition of myoblasts to terminally
differentiated myotubes. Non-coordinate behaviour is not
observed however, following the inhibition of DNA synthesis.

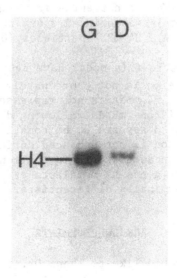

Figure 3. The level of poly(A)$^+$H4 mRNA in growing
myoblasts (G) and in differentiated myotubes (D). Poly(A)$^+$
RNAs were analysed (5ug/lane) as described (Fig. 1).

CONCLUSIONS

We have characterized the behaviour of the multiple subspecies of core histone mRNA during myoblast growth and during differentiation to myotubes. Through a comparison of the half-lives of the various subspecies and their coordinate behaviour we have described the pattern of regulation of one set of genes occurring as proliferating cells withdraw from the cell cycle and differentiate.

Most members of the population of histone mRNA subspecies are coordinately regulated during myoblast differentiation and following inhibition of DNA synthesis. An active and rapid post-transcriptional mechanism seems to exist which coordinately destroys histone mRNA effectively controlling its translation. It is possible that excess histone protein could feed back and activate such a mechanism during periods when DNA synthesis is inhibited through external means or through terminal differentiation. We are currently investigating the nature of the proteins associated specifically with the poly(A)$^+$ and poly(A)$^-$ H4 mRNP particles (6).

Two exceptions to this model have been detected in which gene expression is not coordinately regulated. The poly(A) containing H4 mRNA is not repressed to the same extent as other histone mRNAs following differentiation. H4 mRNA subspecies 3 appears to be constitutively expressed and its accumulation is not enhanced during S phase. Thus a degree of independance exists in the control of the individual histone genes as myoblasts proceed through the cell cycle and terminally differentiate.

ACKNOWLEDGEMENTS

The authors would like to thank Dr. G. Stein for the gift of the human histone clones. We also acknowledge support from the Medical Research Council, Muscular Dystrophy Association and National Cancer Institute of Canada. R.C.B. was supported by a MRC post-doctoral fellowship and F.A.J. was supported by a MDA pre-doctoral fellowship.

REFERENCES

1. Bird, R.C., F. A. Jacobs, G. Stein, J. Stein, and B. H. Sells. 1984. Coordinate regulation of histone mRNAs during growth and differentiation of rat myoblasts. In preparation.
2. Caravatti, M., A. Minty, B. Robert, D. Montarras, A. Weydert, A. Cohen, P. Daubas, and M. Buckingham. 1982. Regulation of muscle gene expression. The accumulation of messenger RNAs coding for muscle-specific proteins during myogenesis in a mouse cell line. J. Mol. Biol. 160:59-76.
3. Holtzer, H., G. Yeoh, N. Rubenstein, J. Chi, S. Fellini, and S. Dienstman. 1977. A review of controversial issues in myogenesis. In Regulation of Cell Proliferation and Differentiation. Plenum Press, NY. 87-104.
4. Nadal-Ginard, B. 1978. Commitment, fusion and biochemical differentiation of a myogenic cell line in the absence of DNA synthesis. Cell 15:855-864.
5. Plumb, M., J. Stein, and G. S. Stein. 1983. Coordinate regulation of multiple histone mRNAs during the cell cycle in HeLa cells. Nucl. Acids Res. 11:2391-2410.
6. Ruzdijic, S.D., R.C. Bird, F.A. Jacobs, and B.H. Sells. 1984. Specific mRNP particles: Characterization of the proteins bound to histone H4 mRNAs isolated from L6 myoblasts. In preparation.
7. Shani, M., D. Zevin-Sonkin, O. Saxel, Y. Carmon, D. Katcoff, U. Nudel, and D. Yaffe. 1981. The correlation between the synthesis of skeletal muscle actin, myosin heavy chain, and myosin light chain and the accumulation of corresponding mRNA sequences during myogenesis. Develop. Biol. 86:483-492.
8. Stein, G.S., and T.W. Borun. 1972. The synthesis of acidic chromosomal proteins during the cell cycle of HeLa S3 cells. I. J. Cell Biol. 52:292-307.
9. Wu, R.S., and W. M. Bonner. 1981. Separation of basal histone synthesis from S-phase histone synthesis in dividing cells. Cell 27:321-330.

THE RECEPTOR BINDING OF INSULIN AND POLY(ADP-RIBOSE)-SYNTHESIS DURING THE CELL CYCLE

H. ALTMANN, O. TÖRÖK[+], P. KOVACS[+],
G. CSABA[+]
Institute of Biology, Research Centre
Seibersdorf, A-2444 Seibersdorf, Austria
[+]Department of Biology, Semmelweis University of Medicine, Budapest, Hungary

ABSTRACT

Hormones can influence the poly(ADP-ribose)-synthesis in the nuclei of cells. But ADP-ribosylation of a component of the receptor adenylate cyclase system influences also cell functions via the cAMP level. In our investigations Chang liver cells, which are target cells for insulin, and HeLa cells were treated with insulin, 3-methoxybenzamide or both agents together and the insulin receptor sites determined with FITC labeled insulin. In HeLa cells insulin as well as methoxybenzamide decreased the FITC-insulin binding to cytoplasmic and nuclear receptor sites. In contrast to this result, both compounds generates higher receptor binding sites in Chang liver cells. Both cell lines were synchronized by thymidine and N_2O treatment. HeLa cells had after pretreatment with insulin, especially during the S-phase, decreased insulin receptor sites. These investigations were done at different times after starting the cell cycle progression after the release of the mitotic block. The highest binding values were obtained at the end of the S-phase and beginning of G_2. The poly-(ADP-ribose)-synthesis in untreated HeLa cells showed 3 peaks. The first peak was at the beginning of G_1, the second at the first part of the S-phase and the third near the end of the DNA syn-

thesis. In insulin pretreated HeLa cells only the
1st and 2nd peak appeared, not the third. The poly-
(ADP-ribose)-synthesis in the 1st part of S-phase
was strongly increased. In Chang liver cells,
insulin pretreatment changes the PAR-synthesis
pattern strongly. From nucleoid sedimentation
studies we could show that poly(ADP-ribose)-synthe-
sis on DNA-nuclear cage binding sites is reduced
after insulin treatment, which is possibly connected to increased transcription activity.

INTRODUCTION

About 30 years ago the understanding of the
separation of the cell cycle in different phases -
G_1, S, G_2 and M was described. The cell division
cycle may be regulated by a programmable clock
within the chromatin and posttranslational modification of proteins play an important role in this
regulation mechanism (1).

Chromatin undergoes transient alterations in
structure during the cell cycle. ADP-ribosylation
seems also to be involved in the regulation of the
phosphorylation of histones, but it has also been
proposed that ADP-ribosylation of a specific protein component of the adenyl cyclase system activates the cAMP production (2). Phosphorylation of
histones by the catalytic subunit of cAMP protein
kinase was reduced when histones were ADP-ribo-
sylated (3). On the other side poly(ADP-ribosyla-
ted) H_1 histones were highly accessible to in vitro
phosphorylation by a nuclear proteinkinase (4).
Certain levels of the phosphorylation of H_1 histones seems to be related to cAMP dependent activation of gene expression. Altered gene expression
and protein kinase activity is also connected to
the insulin function in the cell. Hormones regulate
the expression of restricted sets of genes in a
tissue specific manner.

However, the exact action mechanism of insulin
and the difference in the regulation of cellular
functions in target and nontarget cells is unknown.

A wide range of different cells show insulin
membrane receptors on the surface of cells, which
relates a variety of metabolic processes via the
protein kinase activity located on the receptor

(5). The insulin receptor consists of a α and a β subunit with molecular weight of 135.000 and 95.000 respectively. The protein kinase activity of the receptor phosphorylates its target proteins on tyrosine residues. It is of interest that also the transforming proteins of several RNA tumor viruses possess this rare protein kinase (6). The β-subunit of the insulin receptor is also self phosphorylated by its own kinase activity at a tyrosine residue and has also an ATP-binding site. Insulin bind to the α-subunit and increases the extent of phosphorylation of the β-subunit of the receptor. Adenyl-cyclases are under positive and negative control of hormones and ADP-ribosylation blocks hormone mediated inhibition of adenyl-cyclase activity (7). On the other side insulin can regulate ADP-ribosylation reactions (8). It is up till now not clear if the insulin receptor complex in non or low target cells regulates the unspecific stimulation of proliferation of cells via the control of adenylate cyclase, generates transmembrane signals, is internalized and degradated in lysosomes (9, 10, 11).

In target cells, like liver or fat cells, the insulin receptor complex is internalized, partly degradated, but insulin can also bind to receptors located in the nuclear envelopes. These receptors have different characteristics from plasma membrane receptors (12, 13). Insulin in low concentration reduces ADP-ribosylation reactions and reduces therefore one inhibitor of RNA-synthesis (14). Insulin activates also the nuclear envelope nucleoside triphosphatase and therefore increases the efflux of mRNA from isolated nuclei to the cytoplasma, and decreases ^{32}P incorporation into the nuclear envelope proteins (15). Our experiments have recently shown that insulin also reduce the synthesis of poly(ADP-ribose) in nucleoids, where a peptide of 10.000 daltons is the main acceptor for ADP-ribosylation (16).

MATERIAL AND METHODS

The synchronization of HeLa and Chang liver cells were done according to the method of P.N. Rao (17). Prior to the synchronization steps, cells

were treated with insulin (10^{-6} m) for 4 h at 37°C.
Thymidine (3.10^{-3} M endconcentration) was added
and incubated for 16 h. After 3.75 h incubation
without thymidine the cells were treated with N_2O
at 5.1-5.2 atm for 8.5 h.

To control the synchrony during the S-phase,
tritiated thymidine (0.1 uCi/ml, 2 Ci/mM) was
added to the petridishes and incubated for additional
30 min. The radioactivity incorporated in DNA was
measured after a PCA precipitation in a liquid
szintillation counter and the specific activity
calculated from the DNA content in the sample.

Poly(ADP-ribose)-synthesis was determined by
treatment of permeabilized cells (hypoosmotic cold
shock) with ^{3}H-NAD (1 uCi NAD adenine-2.8-^{3}H, 3 Ci/
mM NEN) (18). The reaction was stopped by cold PCA
and radioactivity was counted in the precipitate.
For the PAR-polymerase inhibition studies 100 uM
solution of 3-methoxybenzamide was used. For
nucleoid sedimentation studies 0.5 million cells
were lysed on the top of a 15-30% sucrose-gradient
using 0.5% Triton X 100, 2 mM Tris, 0.1 M EDTA and
1 M NaCl pH 7.8. Centrifugation was done in a
Beckman L5 ultracentrifuge using a SW 40 rotor for
45 min at 70.000 x g. The gradient was analysed by
a flow photometer at 254 nm and fractionated by a
fraction collector. The radioactivity was determined in each fraction by liquid szintillation
counting.

RESULTS AND DISCUSSION

Figure 1 and 2 show the influence of insulin and
3-methoxybenzamide (MBA) on high target (Chang
liver) and low target (HeLa)cells.

It is of interest that pretreatment with insulin
and MBA result in both cases in decreased FITC-
labeled insulin binding in HeLa cells. It could be
that insulin pretreatment of cells blocks partly
the receptor-binding sites in HeLa cells. But
insulin in low concentration (like MBA) inhibits
also PAR-synthesis (8).

Further studies should show if for the association - dissoziation reaction of hormones with the
receptor, PAR-synthesis or ADP-ribosylation is
necessary. In Chang liver cells both the nuclear

Figure 1

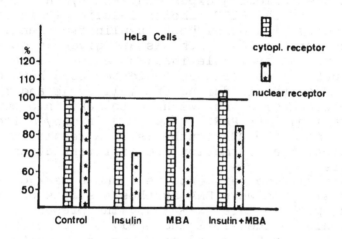

Influence of insulin or 3-methoxybenzamide treatment on HeLa
and chang liver cell insulin receptor binding sites

Figure 2

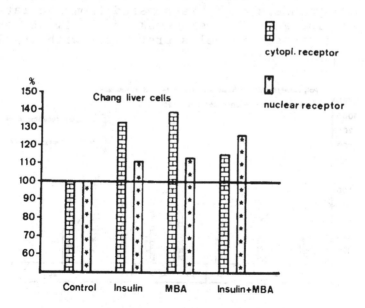

and the cytoplasmic receptor are able to bind more
FITC labeled insulin after the different pretreat-
ments. Additional to some interaction of PAR with
receptors the possibility should not be excluded
that new insulin receptors are synthesized after
the pretreatment.

In the following experiment PAR-synthesis was
compared to the FITC labeled insulin binding in
HeLa cells with or without insulin treatment. The
values for the PAR-synthesis are given as cpm/µg
DNA, and the available insulin receptor sites are
% values. No correlation could be found between
the PAR-synthesis during the cell cycle and the
receptor binding. But we have to take into consideration, that PAR is involved in many different
reactions, like DNA synthesis, DNA repair, transcription, transformation and differentiation (19)
(figure 3).

In earlier experiments we found a peak of PAR-
synthesis just when DNA synthesis starts and a
second peak in late G_2+M (20). The present study
shows high values in M and early G_1, a second
peak in early S and end of S. Generally PAR-synthesis is low when DNA synthesis is high, but for
initiation and termination of DNA synthesis PAR
seems to be necessary. After insulin pretreatment,
the PAR-synthesis peak disappeared from the late
S-phase. The PAR-synthesis peak in the first part
of S is increased in cells pretreated with insulin.

Figure 3

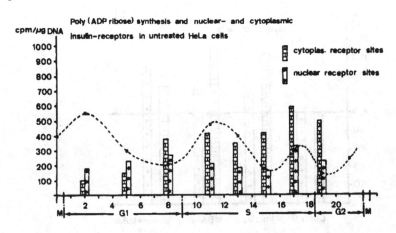

Poly (ADP ribose) synthesis and nuclear- and cytoplasmic
insulin-receptors in untreated HeLa cells

Figure 4

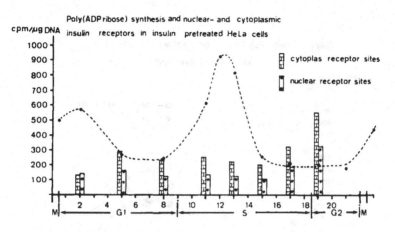

In HeLa chromatin, the major protein acceptor of
ADP-ribose ist PAR-polymerase itself (figure 4).
PAR-polymerase activity was high in extended forms
of chromatin. The level of ADP-ribosylation of
metaphase cells was found to be higher than interphase cells (21). In our experiments all ADP-ribose
found in chromatin can be detected as protein con-
jugates. Radioactivity from NAD-^{3}H was found also
in the nucleoids besides in the self-ADP-ribosy-
lated PAR-polymerase. The radioactivity of PAR or
ADP-ribose bound to the residual proteins of
nucleoids are in all experiments about 15% of the
whole radioactivity of cells, which can be reduced by insulin treatment. Inhibition of PAR-
polymerase by insulin seems to stimulate hormone
dependent transcription in HeLa cells. Chang liver
cells behave different in this respect. FITC-labeled
insulin binding was lower in these pretreated cells
compared to control cells. The highest receptor
binding for nuclear as well as for cytoplasmic
receptors was in the late S-phase and in G_2.
    For comparison of low target with target cells,
PAR-synthesis was also investigated in Chang liver
cells during the cell cycle (figure 5 and 6).
    Pretreatment with insulin changed the PAR-
synthesis during the cell cycle dramatically. Since

Figure 5

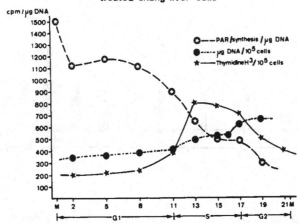

Poly(ADPribose)synthesis during the cell cycle of untreated chang liver cells

Figure 6

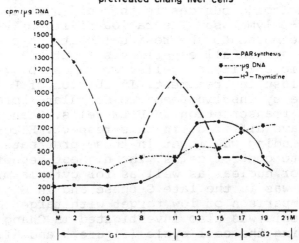

Poly(ADPribose)synthesis during the cell cycle of insulin
pretreated chang liver cells

ADP-ribosylation is also involved in the regulation
of activities of various enzymes as well as structural proteins, the action of insulin on PAR-synthesis in cells is one of the regulatory mechanisms
in cells by which hormones are also acting.
    Studies in our laboratories are in progress to
examine whether such changes in PAR-synthesis by
hormone treatment can influence also differentiation
and transformation processes.

REFERENCES

1. EDMUNDS, L.N, K.J. ADAMA, 1981, Clocked cell
    cycle clocks. Science 211, 1002-1013.
2. KATADA, T., M. UI, 1982, Direct modification of
    the membrane adenylate cyclase system by islet-
    activating protein due to ADP-ribosylation of
    a membrane protein. Proc.Natl.Acad.Sci.USA 79,
    3129-3133.
3. TANIGAWA, Y., M. TSUCHIYA, Y. IMAI, M. SHIMOYAMA,
    1983, ADP-ribosylation regulates the phosphory-
    lation of histones by the catalytic subunit of
    cyclic AMP dependent protein kinase. FEBS
    letters 160, 217-220.
4. WONG, M., M. MIWA, T. SUGIMURA, S. SMULSON,
    1983, Relationship between histone H_1 poly
    (adenosine diphosphate ribosylation) and H_1
    phosphorylation using anti-poly(adenosine
    diphosphate ribose) antibody. Biochemistry
    22, 2384-2389.
5. ROTH, R.A., D.J. CASSELL, 1983, Insulin recep-
    tor: Evidence that it is a protein kinase.
    Science 219, 299-301.
6. INGEBRITSEN, T.S., P. COHEN, 1983, Protein
    phosphatase: Properties and role in cellular
    regulation. Science 221, 331-338.
7. HILDEBRANDT, J.D., R.D. SEKURA, J. CODINA, R.
    IYENGAR, Ch.R. MANCLARK, L. BIRNBAUMER, 1983,
    Stimulation and inhibition of adenyl cyclases
    mediated by distinct regulatory proteins.
    Nature 302, 706-709.
8. TÖRÖK, O., H. ALTMANN, 1982, The influence of
    insulin on poly(ADP-ribose)-synthese and the
    phases of the cell cycle. In: "DNA-Repair and
    Chromatin". Proc.Symp.Saalfelden 1981. Eds.:

H. ALTMANN, G. Klein. Seibersdorf Press, 15-20.

9. SEALS, J.R., L. JARETT, 1980, Activation of
    pyruvate dehydrogenase by direct addition of
    insulin to an isolated plasma membrane/mito-
    chondria mixture: Evidence for generation of
    insulin's second messenger in a subcellular
    system. Proc.Natl.Acad.Sci.USA 77, 77-81.

10. BEACHY, J.C., D. GOLDMAN, M.P. CZECH, 1981,
    Lectins activate lymphocyte pyruvate dehydro-
    genase by a mechanism sensitive to protease
    inhibitors. Proc.Natl.Acad.Sci.USA 78, 6256-
    6260.

11. vanOBERGHEN, E., P.M. SPOONER, C.R. KAHN, S.S.
    CHERNICK, M.M. GARRISON, F.A. KARLSSON, C.
    GRUNFELD, 1979, Insulin-receptor antibodies
    mimic a late insulin effect. Nature 280, 500-
    502.

12. VIGNERI, R., I.D. GOLDFINE, K.Y. WONG, G.J.
    SMITH, V. PEZZINO, 1978, The nuclear Envelope.
    The major site of insulin binding in rat liver
    nuclei. J.Biol.Chem. 253, 2098-2103.

13. HORVAT, A., E. LI, P.G. KATSOYANNIS, 1975,
    Cellular binding sites for insulin in rat
    liver. Biochim.et Biophys.Acta 382, 609-620.

14. SLATTERY, E., J.D. DIGNAM, T. MATSUI, R.G.
    ROEDER, 1983, Purification and analysis of a
    factor which suppresses nick induced trans-
    cription by RNA polymerase II and its identity
    with poly(ADP-ribose)polymerase. J.Biol.Chem.
    258, 5955-5959.

15. PURRELLO, F., R. VIGNERI, G.A. CLAWSON, I.D.
    GOLDFINE, 1982, Insulin stimulation of nucleo-
    side triphosphatase activity in isolated
    nuclear envelopes. Science 216, 1005-1007.

16. BRKIC, G., A. TOPALOGLOU, H. ALTMANN, 1984,
    Poly(ADP-ribose) in nucleoids. OEFZS Ber.Nr.
    4267, BL-450.

17. RAO, P.N., 1968, Mitotic synchrony in mammalian
    cells treated with nitrons oxide at high
    pressure. Science 160, 774-776.

18. ALTMANN, H., I. DOLEJS, 1982, Poly(ADP-ribose)-
    synthesis, chromatin structure and DNA repair
    in cells of patients with different diseases.
    Progr.in Mut.Res. 4, 167-175.

19. ALTMANN, H., 1983, Poly(ADP-ribose)-Synthese
    und Regulationsstörungen bei Erkrankungen.

Wr.klin.Wochenschr. 95, 861-864.

20. ALTMANN, H., 1983, Faktoren, die DNA-Reparatur-
    prozesse innerhalb des Zellzyklus beeinflussen.
    Acta histochem. 27, 87-94.

21. SONG, M.H., K.W. ADOLPH, 1983, ADP-ribosylation
    of nonhistone proteins during the HeLa cell
    cycle. Biochem.Biophys.Res.Comm. 115, 938-945.

A NOVEL METHOD OF TRANSLATION IN FIBROIN

G.C. Candelas, N. Ortiz, A. Ortiz, T.M.
Candelas and O.M. Rodríguez
Department of Biology, University of Puerto
Rico, Rio Piedras, Puerto Rico 00931

A fundamental consideration of any research is that
no single system is expected to ideally or even usefully
reveal information in all aspects of a problem. When the
problem is as complex as the synthesis of proteins and
its regulation, this is all the more evident. It, therefore, follows that a wide variety of experimental systems
is of prime importance, since this, more than the insight
of the investigator, determines what can and cannot be
achieved.

The so called "ideal" systems for these types of
studies have been those with highly differentiated cells,
which at some point in their life become specialized for
the production of one or few specific protein products.
These seem to be relatively simple systems, at least with
respect to the quantification of synthesis and the accumulation of the product and their mRNA templates.

The large ampullate glands of the spider, Nephila
clavipes have well established qualifications as a fruitful model system for these types of investigations. These
highly differentiated structures produce large quantities
of their tissue-specific product during the entire life of
the female adult.

Nephila clavipes is a large spider of wide distribution in the tropical and subtropical areas of the western
hemisphere (1). Within our geographical area, it is very
abundant and easily collected during most part of the

year. They fare well in the laboratory requiring high
moisture.

Orb-Web building spiders produce a series of natural
fibers which are used for the web, dragline, egg sac, or
in swathing their prey. These fibers are synthesized by
five to seven pairs of specialized glands located in the
animals abdominal cavity. It has been reported that each
type of gland secretes one protein for one or two of the
functions previously mentioned (2, 3).

Of these glands, the most prominent are the large
ampullate pair. They comprise from 3-5% of the animal's
wet weight and are capable of producing protein equivalent
to 10% of the glands weight every web building cycle (3).

We have, thus far, isolated the glands' product from
the lumen and analyzed it by SDS-PAGE. These analyses
have confirmed that the gland is a one protein system,
since it produces a single fibroin of approximately
320,000 daltons molecular weight (4).

We succeeded in maintaining the excised glands
metabolically active, for at least four hours, in a simple
culture medium. Under these conditions, they produce the
full size tissue-specific product if properly stimulated.

A special feature of this system is that we are able
to turn the synthesizing activity of the glands on or off
at our will. We have developed a technique that allows
the simultaneous stimulation of a number of organisms
through the mechanical depletion of the organism's stored
silks (5).

Fibroin synthesis can be monitored in the cultured
glands through labeling with tritiated alanine and
glycine which account for approximately 60% of the amino
acid sequence of this fibroin (6, 7).

Excised glands from stimulated and unstimulated
organisms were subjected to time course studies, with short
pulses of labeled amino acids at selected time intervals
after the stimulation. A dramatic and transient wave of
protein synthesis was observed peaking after 90 minutes.
Using pulses of tritiated uridine revealed a similar
response in mRNA synthesis preceeding that of protein by
60 minutes (5). Interestingly, this falls within the
average time required for the processing and translation
of a eukaryotic primary transcript.

The luminar product of these cultured glands turned
out to be the full-size fibroin. Intriguing was a step

ladder array of peptides obtained from extracts of the
secretory epithelium and which became labeled in the
stimulated glands only (figure 1).

Figure 1

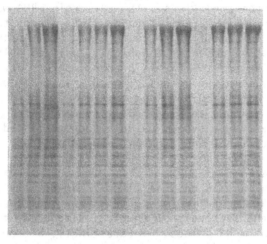

Peptides from secretory epithelium of stimulated glands.

This step ladder of polypeptides, culminating in the final
tissue product, displayed a constant pattern with respect
to the peptides in the gel's lanes and reproduced faith-
fully from one gland preparation to the other. Although
the sites of these bands (reflecting size) were highly
reproducible, their relative intensities (reflecting
length of pause) varied (5).

We were able to establish an exclusive relationship
between these peptides and the process of fibroin synthesis through the labeling kinetics of a series of amino
acids. What we found was that the degree of incorporation of the amino acids correlates with their abundance
in the large ampullate fibroin. This definetely established that the appearance of the peptides was related to
the production of fibroin (8). We concluded that these

peptides must be produced by pauses during the elongation of fibroin such as those reported during the synthesis of silkworm fibroin in intact glands by Lizardi and coworkers (9).

In order to scrutinize the translational mechanism within controllable variables we turned to the process under cell-free conditions. Thus, fibroin mRNA was isolated from stimulated large ampullate glands through a high yield extraction procedure (10). This provided us with preparative amounts of a highly purified poly A+ fraction. The purity of this preparation was attested by its mobility in highly denaturing gels as a homogeneous band approximately 50-60S in size.

The fraction's template activity was tested in a reticulocyte lysate system using either tritiated glycine or alanine to label the translational products. These were analyzed by SDS-PAGE and visualized by fluorography (11) as seen in figure 2.

Using a labeled natural fibroin marker, the gels revealed that the reticulocyte lysate was not capable of supporting the synthesis of the full size product, even under optimal mRNA values. Interestingly, although the full size product did not appear, a step ladder of smaller polypeptides did show up in these gels. These peptides increased in size and also in intensity as the mRNA reached its optional value. The full-size protein product was obtained only when the incubations were supplemented with tRNA extracted from the glands (homologous tRNA) at optimal values. Here, we were able to see a complete ladder of labeled polypeptides culminating in the full size product (figure 2b). This polypeptide array was compared to that obtained in the intact glands and found to display the same distribution of sites along the gel's lanes. The step ladder of nascent fibroin chains, and the absolute requirement for the tRNA supplementation, so as to obtain the fully polymerized fibroin, is also characteristic of the cell-free translation of Bombyx silk mRNA, as observed by two independent groups of investigators (9, 11). We have further characterized our system and found a concentration dependency of translation efficiency on the tRNA supplementation until an optimal value is reached and beyond which a negative response is obtained. As far as our system is concerned the translational pauses are produced in the optimal concentrations

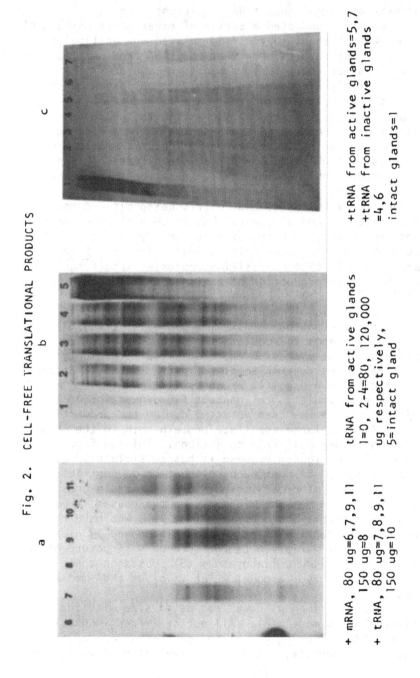

Fig. 2.   CELL-FREE TRANSLATIONAL PRODUCTS

a

+ mRNA, 80 ug=6,7,9,11
        150 ug=8
+ tRNA, 80 ug=7,8,9,11
        150 ug=10

b

tRNA from active glands
1=0, 2-4=80, 120,000
ug respectively,
5=intact gland

c

+tRNA from active glands=5,7
+tRNA from inactive glands
=4,6
intact glands=1

of mRNA and homologous tRNA supplementation.

We have conducted a series of experiments using tRNA
from two different sources: stimulated and unstimulated
glands. Our results show that the tRNA from unstimulated
glands is not equipped to support the synthesis.
Interestingly, it compares to the support given by the
unsupplemented reticulocyte lysate (figure 2, b and c).
Thus, it seems that the applied stimulus, which results
in the enhancement of fibroin synthesis, may also be
maneuvering a shift in the population of tRNA isoacceptors
such as that which occurs during the differentiation for
silk production in Bombyx (13). Here the shift achieves
an adaptation of the glands complement of tRNA
isoacceptors to the codons of tissue-specific product.
This protein, and also ours, has an unusual amino acid
sequence where, in both, alanine and glycine account for
about 60% of the total amino acids(6, 7). The Bombyx
adaptation in tRNA composition expresses itself further
by the timely production of a tissue specific alanine
tRNA (14).

Translational modulation by tRNA has been best
characterized in the Bombyx silk glands (15, 16, 17),
however it seems to modulate the synthesis of other
proteins. The mechanism expresses itself in highly
differentiated systems involved in the synthesis of one
or a few tissue-specific products (18).

The accumulation of nascent fibroin chains of
discrete sizes has been detected during the production of
the full size large ampullate protein both in the intact
gland an under cell-free conditions. This we find
extremely intriguing because it has possible biological
implications. Discontinuous elongation of peptides has
been found in other systems beside the fibroin ones
(19- 23). The mechanism underlying the pauses is yet
unclear. Morris and coworkers (19, 20), who have reported
non-uniform peptide sizes during both α and β globin,
attribute the variation to the secondary structure of the
mRNA. Lizardi and collaborators (9), based on their experimental data, favor a model which involves tRNA-mediated
modulation.

Our experimental data lead us to side with the
Lizardi group model (8). All our cell-free incubations
were conducted using equivalent concentrations of the same
identical template, yet the relative intensities of the

of the bands containing the discrete size nascent fibroin
chains varied in accordance with the nature and/or
concentration of the tRNA supplements. Hence, tRNA
modulates the accumulation of these peptides during the
cell-free translation of fibroin templates.

The fact that translational pauses occur during
fibroin synthesis in intact glands provokes speculation
on its possible biological implication. As far as the
synthesis of fibroin is concerned, and in agreement with
Lizardi et al (9), we see an advantage in adjusting the
elongation rate during the synthesis of these types of
molecules. This might result in the maintenance of
certain levels of polysomes loading and/or adjust the
time during which the proteins post-translational modifi-
cating agent may have access to the nascent chains.

Acknowledgemnt

We wish to acknowledge the fine technical assistance
of Miguel Hernández and José Rodríguez. We appreciate
Mr. and Mrs. Henry Klumb's cooperation by permitting us
to collect on their grounds.

Supported by NSF Grant PCM 81-03284 to Graciela C.
Candelas, Institutional Funds from U.P.R. and a Grant
from U.P.R. Resource Center for Science and Engineering
to Olga M. Rodriguez.

References

1. Moore, C. 1977. Am. Mid. Natur. 98:95-108.
2. Peters, V.H.M. 1955. Z. Naturforsch. 10b:395-404.
3. Peakall, D. 1966. Comp. Biochem. Physiol. 19:253-258.
4. Candelas, G.C. and Cintrón, J. 1981. J. Exp. Zool. 212:
   1-6.
5. Candelas, G.C. and López, F. 1983. Comp. Biochem.
   Physiol. 74B:637-641.
6. Lucas, F. Shaw, J.T. and SMith, S.G. 1960. J. Mol.
   Biol. 2:339-349.
7. Anderson, S.O. 1970. Comp. Biochem. Physiol. 35:705-
   711.
8. Candelas, G.C., Candelas, T. and Ortiz, A. 1983.
   Biochem. Biophys. Res. Comm. 116:1033-1038
9. Lizardi, P., Mahdavi, V., Shields, D. and Candelas,
   G.C. 1969. Proc. Natl. Acad. Sci. USA 76:6311-6215.

10. Lizardi, P.M. and Engleberg, A. 1970. Anal. Biochem.
    98:116-122.
11. Chavancy, G., Marbaix, G., Huez, G. and Cleuter, J.
    1981. Biochimie 63:611-618.
12. Candelas, G.C. and Ortiz, N. (submitted).
13. Garel, J.P., Mandel, P., Chavancy, G. and Daille, J.
    1970. FEBS Lett. 7:327-329.
14. Meza, L. Araya, A. Leon, G. Kraus Kopf, M., Sidiqui,
    M.A.Q. and Garel, J.P. 1977. FEBS Lett. 77:255-
    260.
15. Garel, J.P. 1976. Nature 260:805-906.
16. Garel, J.P., Garber, R.L. and Sidiqui, M.A.Q. 1977.
    Biochem. 16:3618-3624.
17. Chavancy, G., Chevalier, A., Fournier, A. and Garel,
    J.P. 1979. Biochimie 61:71-78.
18. Garel, J.P. 1974. J. Theorte. Biol. 43:211-225.
19. Protzel, A. and Morris, A.J. 1974. J. Biol. Chem.
    249:4594-4600.
20. Chaney, W. and Morris, A.J. 1979. Arch. Biochem.
    Biophys. 194: 283-291.
21. von Heijne, G., Nilsson, L. and Blomberg, C. 1978.
    Eur. J. Biochem. 92:397-402.
22. Abraham, A. and Pihl, A. 1980. Eur. J. Biochem. 106:
    257-262.
23. Randall, L., Josefsson, L.G. and Hardy, S.J. 1980.
    Eur. J. Biochem. 107:375-379.

TUBULIN AND ACTIN GENE EXPRESSION DURING THE CELL CYCLE

S. Zimmerman, A.M. Zimmerman, J. Thomas and
I. Ginzburg.
Division of Natural Science, Glendon College,
York University, Toronto, Ontario; Dept. of
Zoology, University of Toronto, Toronto, Ont.;
Oncology Research Group, University of Calgary,
Calgary, Alberta and the Dept. of Neurobiology,
The Weizmann Institute of Science, Rehovot,
Israel.

Tubulin and actin are cytoskeletal proteins that play
a central role in the structure and function of dividing
cells (6,10). The control of tubulin and actin gene expression may, therefore, be related to specific cell cycle
events. Tetrahymena pyriformis, a ciliated protozoan, has
been used extensively for cell cycle studies because a high
degree of division synchrony may be induced in this system
(18). Previous investigations of tubulin synthesis regulation, in division synchronized Tetrahymena, suggest that
the initiation of tubulin synthesis depends upon mRNA synthesis (2,3,4). However, these studies do not resolve the
question of whether the regulation of tubulin synthesis
exists at the messenger RNA transcriptional level or at
post transcriptional stages. The presence of actin in
Tetrahymena has been a matter of recent controversy (11,13,
14,17). In the present study, we investigated tubulin and
actin gene expression during the cell cycle. Cloned cDNA
probes for tubulin and actin were used to identify tubulin
and actin mRNA (7,8,9,15). In addition, we examined mRNA
during the synchronous cell cycle in a cell free translation system derived from reticulocytes.

Log growth cultures of Tetrahymena pyriformis are
division synchronized (18) by the administration of a series
of seven 30 minute heat shocks, each spaced one cell gener-

107

ation apart (157 min). Following the end of the last heat
shock (designated EH) the cells progress through G₂ phase
of the cell cycle and undergo synchronous division at

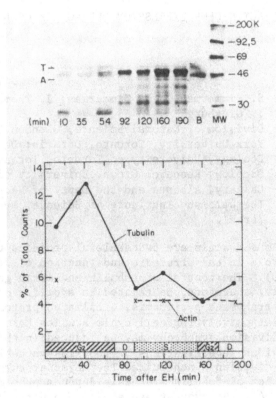

Figure 1 Analysis of translation products in a reticulocyte cell free translation system directed by poly (A⁺)
mRNA sequences,derived from division synchronized Tetrahymena. At specific times after the last heat shock (designated EH) polysomes were isolated from cells. Poly (A⁺)
mRNA was separated from the polysomes on an oligo (dT) cellulose column and translated in a cell free reticulocyte
system. The fluorogram of the separated ³⁵S methionine
labelled proteins are shown. Densitometric trace of the
fluorogram indicate the amount of tubulin and actin synthesized. The numbers below each lane represent the time EH.
Material in lane B are products directed by rat brain poly
(A⁺) mRNA. Molecular weight markers are shown in the far
right lane. Tubulin (T) and actin (A) are shown.

approximately 85 and 200 minutes EH. The number of cells
doubles after each synchronous division. The interval
between the first and second synchronous division is designated the free-running cell cycle and is considered to
be similar to the non-induced log growth cell cycle.

   Polysomes were isolated by sucrose gradient centrifugation from synchronized Tetrahymena at specific cell cycle
stages. Poly (A$^+$) containing mRNA was separated from the
polysomes by oligo (dT) cellulose chromatography and translated in a rabbit reticulocyte lysate system containing ^{35}S
methionine (16). The cell free products were analyzed by
SDS polyacrylamide gel electrophoresis using linear and
gradient gels. The fluorogram of separated ^{35}S methionine
labelled proteins directed by mRNA is shown in Figure 1.
Tubulin and putative actin were identified in the translation products. Both proteins were found to comigrate with
actin and tubulin standards which were translated from rat
brain poly (A+) mRNA. Quantitative analyses of the ^{35}S
labelled proteins were performed by densitometric scanning
of the fluorogram. Tubulin was present in all the cell free
products synthesized with mRNA preparations isolated from
cells preceding the first synchronous cell division and
during the free-running cell cycle. Tubulin synthesis rises
to a peak value before the first synchronous division (during G$_2$) and falls as the cell progresses through division.
During the free-running cell cycle, message directed tubulin
synthesis fluctuates. Actin was identified as the 45 kDa
band seen during the free running cell cycle. Densitometric
scans of the gel show that the amount of material in the 45
kDa band does not vary as the cell progresses through the
cell cycle.

   In order to identify tubulin and actin mRNA species,
poly (A+) mRNA from division synchronized cells, at different stages of the cell cycle, was fractionated on denatur-
ating agarose-formamide gels. The fractionated poly (A+)
mRNA was transferred to nitrocellulose filters and hybridized to nick-translated cDNA probes from plasmids pT 25 and
pA 72 containing tubulin and actin DNA sequences, respectively. Tubulin message sequences rise from a low value
immediately after the last heat shock to a maximum level
around 85 min when the first synchronous division occurs;
at this time, tubulin mRNA increases around 13 fold. During
the free-running cell cycle, a low tubulin mRNA value is

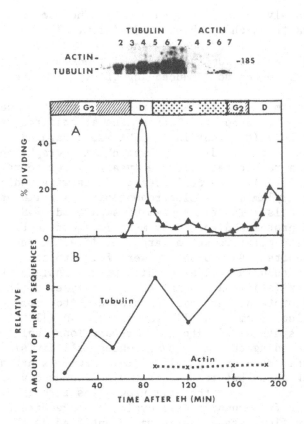

Figure 2   Fractionation of poly (A$^+$) containing mRNA by
agarose-formamide gel electrophoresis and analysis of the
hybridization with tubulin and actin cDNA probes during the
synchronous cell cycle of <u>Tetrahymena</u>. A) <u>Tetrahymena</u> were
synchronized by the one heat shock per generation method of
Zeuthen (18). The time EH represents the time after the
last thermal shock. B) Poly (A$^+$) mRNA (5-10 ug) at specified
times during the cell cycle was fractionated on 1.1% agarose
formamide slab gel. Fractionated RNA was transferred to
nitrocellulose by blotting filters and hybridized with nick
translated tubulin cDNA probe (pT25) and actin cDNA probe
(pA72). A fluorogram of the nitrocellulose filter with
tubulin and actin probes is shown at the top of the figure.
Quantitative determination of the quantity of hybridized
mRNA was performed by densitometric scanning of the fluoro-
grams. The 18S marker and specific cell stages, G$_2$ (Gap 2),
D (division) and S (synthesis) are shown.

reached during the S phase; the mRNA level rises at G_2 and
is maintained at the second synchronous division. The
fluctuations in the fractionated message sequences are
similar to those found in hybridization studies with non-
fractionated poly (A+) mRNA, which was hybridized on nitrocellulose discs to labelled cDNA probes derived from
plasmids pT25. In Tetrahymena, tubulin synthesis is
thought to be regulated by changes in the size of the soluble tubulin pool. Induction of tubulin during G_2 occurs
just following formation of the oral apparatus which requires a large quantity of tubulin. Comparison of in vivo
tubulin synthesis (2) with the induction of tubulin mRNA
sequences during the free-running cell cycle reveals a
close temporal relationship. The absence of a lag phase
provides evidence for transcriptional control of tubulin
synthesis. Furthermore, if tubulin synthesis were under
post-transcriptional control one would expect a constant
synthesis of message. We propose that pool depletion, during the cell cycle, initiates transcription of tubulin
sequences. Fluctuations in the size of the tubulin monomer
pool are reported to exert a modulating influence on tubulin
synthesis in several organisms. Moreover, the mRNA dependent
changes in the level of tubulin appear to be controlled by
the size of the pool of unpolymerized tubulin (1,3,5). The
fluctuation of tubulin mRNA suggests that new species of
isotubulin forms may be required in preparation for cytokinesis.

The cDNA actin probe pA72 plasmids also hybridize with
Tetrahymena fractionated poly (A+) mRNA at a band position
which corresponds to the migration of poly (A+) mRNA species
coding for actin. Actin mRNA is readily seen during the
free-running cell cycle (Figure 2) but not clearly identified during the period preceding the first synchronous
division. It does not show periodicity and its levels are
about 20% of the tubulin mRNA. The identification of actin
mRNA in division synchronized Tetrahymena by hybridization
with cDNA actin probe supports and extends other studies of
actin in log growth Tetrahymena. In these studies (11,12,
13) actin was identified by co-migration of Tetrahymena
actin with rabbit muscle actin, peptide mapping with S.
aureus V8 protease and chymotrypsin, DNase I chromotography
and reactivity with an anti-actin antiserum as determined
by an enzyme-linked immunosorbent assay providing further
strong supportive evidence that Tetrahymena possess actin.

The absence of mRNA fluctuation does not preclude a role
for actin in cell division but may be indicative of a
different regulatory mechanism from that of tubulin.

The regulation of synthesis of proteins during the cell
cycle depends upon the availability of specific mRNA. This
regulation may in turn result from the level of transcription, translation or the stability of the given mRNA molecule. Our hybridization results show fluctuations in the
levels of tubulin mRNA sequences during the cell cycle.
These changes appear to result from changes in the transcription of tubulin genes rather than from a change in the
efficiency of translation or change in the stability of
tubulin mRNA. The similarity of in vivo tubulin synthesis
previously reported (2) and tubulin mRNA levels determined
in the current study is strong support for transcriptional
control. Another possibility for regulation of tubulin
synthesis, namely, through a change in the half life of
mRNA cannot be excluded. However, previous work (1,5) has
shown short lived tubulin mRNA of about two hours, whereas
the time period of our fluctuations was shorter (about 30
min). Thus we suggest that tubulin synthesis is under
direct transcriptional control and its rate of synthesis
may be correlated with its functional role during the cell
cycle.

Acknowledgements

This work was supported in part by grants from BSF
(2923/82) to I.G. and from NSERC (Canada) to A.M.Z. This
work was conducted while A.M.Z. and S.Z. were visiting
scientists at the Weizmann Institute of Science, Rehovot,
Israel. The authors are grateful to Professor U.Z. Littauer
for his continuous interest during the course of this work.

References

1. Ben-Ze'ev, A., Farmers, S.R. and S. Penman. 1979.
    Mechanisms of regulating tubulin synthesis. Cell 17:319-325.
2. Bird, R.C. and A.M. Zimmerman. 1981. Tubulin synthesis
    during the synchronous cell cycle of Tetrahymena.
    Can. J. Biochem. 59:937-943.
3. Bird, R.C. and A.M. Zimmerman. 1984. Abundance of
    tubulin mRNA on polysomes following deciliation and
    during the synchronous cell cycle of Tetrahymena. Cell

and Tissue Kinet. In press.

4. Bird, R.C., Zimmerman, S. and A.M. Zimmerman. 1980. In: Nuclear-Cytoplasmic Interactions in the Cell Cycle (ed. Whitson, G.) pp. 204-221. Academic Press, New York.

5. Cleveland, D.W., Lopata, M.A., Sherline, P. and M.W. Kirschner. 1981. Unpolymerized tubulin modulates the level of tubulin mRNAs. Cell 25:537-546.

6. Dustin, P. 1978. Microtubules, Springer-Verlag, Berlin.

7. Fellous, A., Ginzburg, I. and U.Z. Littauer. 1982. Modulation of tubulin mRNA levels by interferon in human lymphoblastoid cells. EMBO Journal 1:835-839.

8. Ginzburg, I., deBaetselier, A., Walker, M.D., Behar, L., Lehrach, H., Frishauf, A.M. and U.Z. Littauer. 1980. Brain tubulin and actin cDNA sequences: Nucleic Acids Res. 8:3553-3564.

9. Ginzburg, I., Behar, L., Givol, D. and U.Z. Littauer. 1981. The nucleotide sequence of rat α-tubulin: 3'-end characteristics, and evolutionary conservation. Nucleic Acids Res. 9:2691-2697.

10. Goldman, R., Pollard, T. and J.L. Rosenbaum. 1976. Cell Motility, Cold Spring Harbor Laboratory, Cold Spring Harbor, New York.

11. Mitchell, E.J. and A.M. Zimmerman. 1982. Characterization of the actin-like protein from Tetrahymena. J. Cell Biol. 95:281a.

12. Mitchell, E.J. and A.M. Zimmerman. 1984. Biochemical evidence for the presence of an actin protein in Tetrahymena pyriformis. J. Cell Sci. (submitted).

13. Mitchell, E.J., Zimmerman, A.M. and A. Forer. 1981. Identification of a protein presumed to be actin from Tetrahymena. J. Cell Biol. 91:308a.

14. Muncy, L.F. and J.S. Wolfe. 1981. Evidence for a non-actin inhibitor of deoxyribonuclease 1 (DNase 1) in Tetrahymena thermophila. J. Cell Biol. 91:303a.

15. Nudel, U., Katcoff, D., Zakut, R., Shani, M., Carmon, Y., Finer, M., Czosnek, H., Ginzburg, I. and D. Yaffe. 1982. Isolation and characterization of rat skeletal muscle and cytoplasmic actin genes. Proc. Natl. Acad. Sci. USA 79:2763-2767.

16. Pelham, H.R.B. and R.J. Jackson. 1976. An efficient RNA-dependent translation system from reticulocyte lysates. Eur. J. Biochem. 67:247-256.

17. Williams, N.E., Vaudaux, P.E. and L. Skriver. 1979.
     Cytoskeletal proteins of the cell surface in Tetra-
     hymena. I. Identification and localization of major
     proteins. Exp. Cell Res. 123:311-320.
18. Zeuthen, E. 1971. Synchrony in Tetrahymena by heat
     shocks spaced a normal cell generation apart. Exp.
     Cell Res. 68:49-60.

SECTION II

GROWTH ACTIVATION AND DORMANCY

REGULATION OF NUCLEAR AND MITOCHONDRIAL RNA PRODUCTION

AFTER SERUM STIMULATION OF QUIESCENT MOUSE CELLS

Dylan R. Edwards and David T. Denhardt
Cancer Research Laboratory
University of Western Ontario
London, Ontario
Canada   N6A 5B7

ABSTRACT

Using cDNA clones of murine RNA, we have found that the proportion of mitochondrial polyA$^+$mRNA relative to nuclear polyA$^+$mRNA in preparations of cytoplasmic polyA$^+$mRNA decreases when cultured secondary mouse embryo fibroblasts proceed out of G_0 and through G_1 as the result of serum stimulation of quiescent cells. However, we were unable to detect a change in the relative rates of transcription of the mitochondrial and nuclear genomes. Our working hypothesis is that there is a preferential enhancement of processing and/or transport of RNA synthesized in the nucleus. Certain other clones in our library, which was enriched for low abundance species, correspond to nuclear-coded mRNA species whose relative abundance changes after serum stimulation.

INTRODUCTION

Stimulation of quiescent, serum-starved murine fibroblasts to re-enter the cell cycle by addition of medium containing an increased level of serum and growth factors causes an increase in (i) the rate of synthesis of rRNA (13) and in (ii) the accumulation of polyA$^+$mRNA in the cytoplasm (8). The result is an approximate doubling of the cytoplasmic mRNA/rRNA ratio by the end of the first cell cycle. We began the work described here several years ago on the premise that in the newly synthesized mRNA popu-

117

lation there would be species that encoded proteins whose
expression was necessary for the progression of the cell
through the cell cycle. Considerable experimental evidence
was consistent with this view (e.g. 14). Furthermore there
have been several reports recently of examples of genomic
and cDNA clones corresponding to mRNAs whose concentration
in the polyA+mRNA population increases after quiescent
cells are stimulated to leave G_0 (3, 4, 10).

RESULTS

A cDNA library was constructed from the cytoplasmic
polyA+mRNA extracted from a subconfluent population of
secondary Swiss mouse embryo fibroblasts by inserting dC-
tailed duplex cDNA molecules into the dG-tailed Pst site of
pBR322. With one exception, this library was made using
procedures generally used in constructions of this sort
(11); the details will be reported elsewhere. The one
exception was that the population of 'first-strand' cDNA
molecules was depleted of abundant species by annealing to
mRNA extracted from a similar population of cells. Among
the clones (about 20%) that could be detected by hybridization to radioactive cDNA preparations, some were found

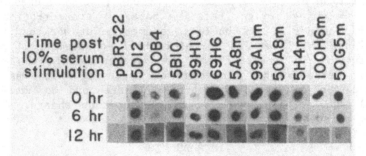

Figure 1: Colony hybridization with [32P]cDNA probes from
quiescent and serum-stimulated cells. PolyA+mRNA from
quiescent cells (0 hour) and from cells 6 and 12 hours
after stimulation with medium containing 10% fetal bovine
serum was used to prepare [32P]cDNA probes. Equivalent
amounts of DNA at the same specific activity (10^8 cpm/μg)
were hybridized to nitrocellulose filter replicas of the
cDNA library (11). The letter m in the clone designations
indicates that the clone hybridized to the mouse mitochondrial DNA genome in plasmid pAMI.

that appeared to correspond to RNA species that increased
or decreased in abundance when quiescent cells were stimulated to enter G_1.   Fig. 1 shows a representative set of
colony hybridizations of selected clones.

Examples of clones whose abundance did not appear to
change (5D12, 100B4) or whose abundance appeared to
increase (5B10, 99H10) are illustrated in Fig. 1; the latter may encode functions that are important for progression
through the cell cycle and will be described in more detail
in subsequent publications. However, it was also apparent
from colony hybridization screening of our cDNA library
that it contained clones that frequently gave decreased
signals with radioactive cDNA probes from stimulated cells
compared to probes from quiescent cells. The majority of
these clones were identified as being of mitochondrial origin by their cross-hybridization to the insert from pAMI,
which contains the entire 16,295 bp mouse mitochondrial
genome (12). Northern blot analyses of polyA+ preparations,
shown in Fig. 2, confirmed this result; in this particular
experiment the apparent cytoplasmic levels of two repre-

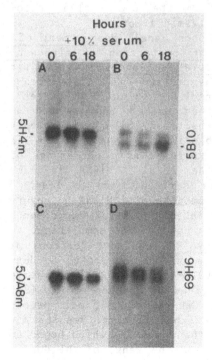

Figure 2:   Northern blot
analysis of polyA+mRNA from
quiescent and serum-stimulated   cells.   Glyoxal-
denatured   polyA+mRNA   (2
µg/lane) from   quiescent
cells   and   from cells 6
hours and 18 hours after
stimulation with serum was
electrophoresed   through
1.1% agarose gels,   then
transferred   to   nitrocellulose by the method of
Thomas (15). The blots were
hybridized with $2x10^6$ cpm
of nick-translated plasmid
DNA   from   the indicated
clones: a, 5H4m; b, 5B10 to
the same blot (a) without
elution of the bound 5H4m
probe; c, 50A8m; d, 69H6 to
the same blot (c) without
elution of the bound 50A8m
probe.

sentative mitochondrial polyA⁺mRNAs had fallen at 18 hours
post-stimulation to approximately one-half their concentration in the quiescent cells.   In contrast, the mRNA
corresponding to 5B10 had increased three to four-fold at

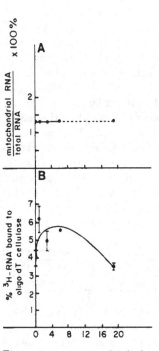

Figure 3:   Relative rates of
synthesis   of   mitochondrial
RNA   and   polyA⁺mRNA   in
quiescent   and   serum-stimulated   cells.   Quiescent
(serum-deprived)   secondary
cultures   of   mouse   embryo
fibroblasts   grown   in   6   cm²
dishes   were   stimulated   with
medium   containing   10%   serum.
The   cultures   were   radio-
labelled   with   1.5   ml   medium
containing   100   µCi/ml   [³H]-
uridine   for   1   hour,   ter-
minating   at   the   times   indicated,   and   RNA   was   extracted
from   the   post-nuclear   cytoplasmic   fraction.

Upper panel: Measurement of
the   proportion   of   mitochondrial   sequences: [³H]RNA   was
hybridized   to   filter-bound
pAM1 DNA (5 µg/20 mm filter),
then   washed   and   counted   as
described   (6).   Radioactivity
binding   to   the   filters   is
expressed   as   a   percentage   of
the   total   [³H]RNA   in   each
hybridization.   Each   point   is   derived   from   at   least   3
hybridizations using different amounts of [³H]RNA to ensure
that the filters were not saturated.   Control experiments
indicated that the hybridization had gone to completion.

Lower panel: Estimation of polyA⁺RNA content: Portions of
the RNA preparations (50,000 cpm of ³H) were bound to dup-
licate 25 mg portions of oligo(dT)-cellulose (Collaborative
Research Inc., Type 3) as described by Johnson et al. (8).
After extensive washing, bound RNA was eluted, acid-preci-
pitated and collected on glass fibre filters. Results are
expressed as the percentage of the input RNA which bound to
oligo(dT)-cellulose; bars indicate the range of the dupli-
cate values.

18 hours after serum stimulation. Clone 69H6, which is not of mitochondrial origin, showed reduced levels in the stimulated cells.

Using a clone found to correspond to mitochondrial 16S rRNA (174C4m), we have shown that this RNA, which is also found in the cytoplasmic polyA⁺mRNA preparations, also followed the same pattern. The reduction in the relative concentration of mitochondrial RNAs was not a result of an actual decrease in the number of RNA molecules per cell; for example, the RNA corresponding to 5H4m was found to accumulate continuously in stimulated cells (data not shown).

We have investigated whether the apparent reduction in abundance of transcripts of mitochondrial origin in the RNA preparation could be due to changes in the rates of synthesis of mitochondrial RNA relative to cellular rRNA and polyA⁺mRNA. Fig. 3A shows that following serum stimulation of quiescent cells the proportion of pulse-labeled RNA which was mitochondrial remained constant relative to total cytoplasmic RNA. It should be emphasized that since the mitochondrial ribosomal RNAs are the most abundant mitochondrial RNA species in cells (2, 5), we have effectively compared the level of these species to the cytoplasmic rRNA in these labelled preparations. The amount of labeled polyA⁺mRNA (nuclear and mitochondrial) in these samples was determined by their binding to oligo(dT)-cellulose. Fig.3B shows that the the proportion of the pulse-labeled [³H]RNA that is able to bind to oligo(dT)-cellulose increases in the first few hours after serum stimulation. These data are consistent with the observations on the rates of accumulation of RNA species in serum-stimulated 3T6 fibroblasts (9).

DISCUSSION

Our data suggest that newly synthesized mitochondrial RNA represents the same constant percentage of total newly synthesized cellular RNA in both quiescent and serum-stimulated mouse fibroblasts. To our knowledge, this fact has not been reported in the literature. If this is so, then why does the proportion of polyA⁺mRNA that represents mitochondrial sequences drop after serum stimulation? We suggest that the apparent reduction in the relative amount of mitochondrial RNA that binds to oligo(dT)-cellulose is

the result of an increased rate of accumulation of nucleus-
derived polyA$^+$mRNA in the cytoplasm in the first few hours
following serum stimulation. Increased cytoplasmic polyA$^+$-
mRNA content is not due to an increase in the rate of hnRNA
transcription (13) but rather represents enhanced processing and transport of polyA$^+$mRNA from the nucleus to the
cytoplasm (9).

Several factors will have to be better defined before
we can fully interpret these observations. For example,
the average length of the polyA tail in HeLa cell mitochondrial mRNA was estimated to be 55 nucleotides (7), somewhat
shorter that the 150-200 nucleotide stretch associated with
cytoplasmic polyA$^+$mRNA of nuclear origin; we don't have any
information on whether there is a change in the average
polyA tail length of mitochondrial polyadenylated transcripts after serum stimulation. This could affect recovery
from the oligo(dT)-cellulose column. The average half-life
of cytoplasmic polyA$^+$ mRNA is approximately 9 hours in both
quiescent and exponentially growing 3T6 cells (1). Mitochondrial mRNAs have been found to be relatively unstable
in HeLa cells, with half-lives varying from 25-90 minutes
(5). Our data could be explained by changes in the stability of mitochondrial mRNAs when quiescent cells are stimulated to divide. Finally since these experiments were performed on RNA from post-mitochondrial supernatants, the
results could be explained by an increased resistance of
mitochondria to lysis by NP-40 during preparation of cytoplasmic extracts from the stimulated cells.   This is
presently under investigation, and preliminary results tend
to discount it as a possibility.

REFERENCES

1.   Abelson, H.T., L.F. Johnson, S. Penman, H. Green.
     1974. Changes in RNA in Relation to Growth of the
     Fibroblast: II. The Lifetime of mRNA, rRNA, and tRNA
     in Resting and Growing Cells. Cell 1: 161-165.

2.   Cantatore, P., M.N. Gadaleta, and C. Saccone. 1984.
     Determination of Some Mitochondrial RNAs Concentration
     in Adult Rat Liver. Biochem. Biophys. Res. Comm. 118:
     284-291.

3.   Cochran, B.H., A.C. Reffel, and C.D. Stiles. 1983.
     Molecular Cloning of Gene Sequences Regulated by

Platelet-derived Growth Factor. Cell 33: 939-947.

4. Foster, D.N., L.J. Schmidt, D.P. Hodgson, H.L. Moses,
    and   M.J.   Getz.   1982.   Polyadenylylated   RNA
    complementary to a mouse retrovirus-like multigene
    family is rapidly and specifically induced by
    epidermal growth factor stimulation of quiescent
    cells. Proc. Natl. Acad. Sci. USA 79: 7317-7321.

5. Gelfand, R., and G. Attardi. 1981. Synthesis and
    Turnover of Mitochondrial Ribonucleic Acid in HeLa
    Cells: the Mature Ribosomal and Messenger Ribonucleic
    Acid Species Are Metabolically Unstable. Mol. Cell.
    Biol. 1: 497-511.

6. Hendrickson, S.L., J-S.R. Wu, and L.F. Johnson. 1980.
    Cell Cycle Regulation of Dihydrofolate Reductase mRNA
    Metabolism in Mouse Fibroblasts. Proc. Natl. Acad.
    Sci. USA 77: 5140-5144

7. Hirsch, M., and S. Penman. 1973. Mitochondrial
    Polyadenylic Acid-containing RNA: Localization and
    Characterization. J. Mol. Biol. 80: 379-391.

8. Johnson, L.F., H.T. Abelson, H. Green, and S. Penman.
    1974. Changes in RNA in Relation to Growth of the
    Fibroblast. I. Amounts of mRNA, rRNA, and tRNA in
    Resting and Growing Cells. Cell 1: 95-100.

9. Johnson, L.F., J.G. Williams, H.T. Abelson, H. Green,
    and S. Penman. 1975. Changes in RNA in Relation to
    Growth of the Fibroblast. III. Posttranscriptional
    Regulation of mRNA Formation in Resting and Growing
    Cells. Cell 4: 69-75.

10. Linzer, D.I.H., and D. Nathans. 1983. Growth-related
    Changes in Specific mRNAs of Cultured Mouse Cells.
    Proc. Natl. Acad. Sci. USA 80: 4271-4275.

11. Maniatis, T., E.F. Fritsch, and J. Sambrook. 1982.
    Molecular Cloning. A Laboratory Manual. Cold Spring
    Harbor Laboratory, Cold Spring Harbor. N.Y.

12. Martens, P.A., and D.A. Clayton. 1979. Mechanism of
    Mitochondrial DNA Replication in Mouse L-cells:
    Localization and Sequence of the Light-strand Origin

of Replication. J. Mol. Biol. 135: 327-351.

13.   Mauck, J.C., and H. Green. 1973. Regulation of RNA
      Synthesis in Fibroblasts During Transition from
      Resting to Growing State. Proc. Natl. Acad. Sci. USA
      10: 2819-2822.

14.   Thomas, G., G. Thomas, and H. Luther. 1981. Transcrip-
      tional and Translational Control of Cytoplasmic Pro-
      teins after Serum Stimulation of Quiescent Swiss 3T3
      cells. Proc. Natl. Acad. Sci. USA 78: 5712-5716.

15.   Thomas, P.S. 1980. Hybridization of Denatured RNA and
      Small DNA Fragments Transferred to Nitrocellulose.
      Proc. Natl. Acad. Sci. USA 77: 5201-5205.

ACKNOWLEDGEMENTS

This research was supported by the National Cancer
Institute of Canada, the Medical Research Council of
Canada, and the Nelson Arthur Hyland Foundation. We thank
Martha Holman for her competent and dedicated technical
assistance and David A. Clayton for the plasmid pAMI. The
manuscript and illustrations were skillfully prepared by
Linda Bonis and Dale Marsh, to whom we extend our grateful
appreciation.

A DECREASE IN THE STEADY STATE LEVEL OF DNA STRAND BREAKS

AS A FACTOR IN THE REGULATION OF LYMPHOCYTE PROLIFERATION

W.L. Greer and J.G. Kaplan

Department of Biochemistry, University of Alberta

Edmonton, Alberta, Canada T6G 2H7

ABSTRACT

Recent reports have shown that resting human and mouse lymphocytes contain DNA strand breaks which must be repaired after mitogen stimulation via a system that is stimulated by ADP ribosylation, before blast transformation and DNA synthesis can occur. We now report that the production of DNA strand breaks in resting cells and their subsequent rejoining after stimulation are not single, unique, punctual events; there is rather a continuous production and repair of strand breaks in both resting and stimulated cells. The equilibrium between strand breakage and repair shifts in the direction of repair following mitogenic activation of splenic lymphocytes. The decrease in strand breaks after stimulation probably results from a transient increase in repair rather than from a decrease in production of breaks. This is consistent with the increase in rate of ADP ribosylation activity in permeabilized cells soon after mitogen treatment. The low level of NAD^+ in resting lymphocytes is a rate-limiting factor in ADP ribosylation, in rejoining of DNA strand breaks and in initiation of the train of events leading to DNA synthesis.

INTRODUCTION

Previous work in this laboratory showed that a 5h pulse with 100 µM 5-fluorouracil irreversibly prevented

125

mitogen-stimulated mouse lymphocytes from entry into DNA
synthesis (1). This treatment produced a great number of
DNA strand breaks in resting or stimulated cells; these
breaks were not repaired, even though lymphocytes have a
very active system for repair of radiation damage (2) and
could account for the inhibition of DNA synthesis by the
pyrimidine analog.

We have recently observed that resting mouse lymphocytes contain inherent DNA strand breaks most of which were
repaired via an ADP ribosylation-dependent mechanism); this
repair was required for cell proliferation. DNA strand
breaks have also been found in other non-proliferating cell
types such as chick myoblasts (3) and human lymphocytes
(4). It has been proposed that these inherent strand
breaks may inhibit cell proliferation and may serve as a
regulatory factor in cell differentiation (4,5).

We now report that an accumulation of DNA strand
breaks occurs when cells are incubated with the benzamide
inhibitors of ADP ribose polymerase. These agents prevent
subsequent entry of treated mouse splenocytes into S phase
even when added to cell cultures many hours after the
number of DNA strand breaks reached its minimum (2h culture). This indicates that there is a continuous production of strand breaks in both resting and stimulated cells
which are normally rejoined via ADP ribosylation. Our data
support the hypothesis that a decrease in the steady state
number of DNA strand breaks occurs after mitogen treatment
and is required for lymphocyte activation. The low cellular level of NAD^+ is a rate-limiting factor in rejoining
DNA strand breaks in resting lymphocytes.

METHODS

Cell Preparation and Culture

Balb/c male mice 8-12 weeks old were killed by cervical dislocation, and their spleens disrupted through a
wire screen. Red blood cells were removed by lysis with
0.83% ammonium chloride. Cells were cultured in RPMI 1640
medium with 10% fetal calf serum, and 2 mM glutamine, at a
density of 2×10^6 cells/ml.

Detection of DNA Strand Breaks

DNA strand breaks were detected using the fluorometric analysis of DNA unwinding technique developed by Birnboim and Jevcak (6). From the percentage of double stranded DNA remaining after alkaline treatment, one can estimate the number of DNA strand breaks from a calibration curve obtained from cells treated with various doses of gamma radiation (7).

Assays for ADP Ribosylation and NAD

ADP ribosylation was measured as incorporation of $[^3H]$ NAD into the acid insoluble fraction of permeabilized cells according to the method of N.A. Berger and Johnson (8).

NAD^+ levels were measured according to the method of Nisselbaum and Green (9).

RESULTS

Table 1 shows that within 2h of onset of concanavalin A (Con A) stimulation of mouse lymphocytes, approx. 3200 DNA strand breaks per diploid genome had been repaired and a minimum level of breaks was reached. Methoxybenzamide (MBA), an inhibitor of ADP ribose polymerase, prevented the Con A-induced repair of these breaks, as well as most of the events of cell stimulation. MBA had a small effect on the increase in activity of the $Na^+ K^+$ ATPase, which is an early essential event required for lymphocyte activation; however protein, RNA and DNA synthesis, as well as blast transformation (all measured at 48h after Con A stimulation) were severely inhibited if MBA was added soon after initiation of culture with Con A.

We and others (3,5) have interpreted the experiments like those of Table 1 to indicate that resting lymphocytes contain DNA strand breaks, and that once repair of these breaks via an ADP ribosylation-dependent mechanism has occurred the cells can undergo blast transformation and DNA synthesis. However, this hypothesis does not fit more recent data. Table 2 shows that even when the inhibitor of repair was added at 26h, which is 24h after strand breaks were at a minimum in mouse, DNA synthesis was inhibited almost as much as when MBA was added at time 0. This suggested to us that DNA strand breaks were continually

Table 1. Effect of methoxybenzamide (MBA) on repair of DNA strand breaks, and other events of lymphocyte activation.

Treatment	A Strand breaks repaired (per diploid genome)				B Events of lymphocyte activation assayed at 48 h				
	Time after addition of Con A				Na⁺ K⁺ ATPase	Blasts	³H Leucine	³H Uridine	³H Thymidine
	0h	1h	2h	3h	μmoles ⁸⁶Rb/ 10⁶ cells/ min	% of total	cpm/10⁶ cells		
Resting	0	0	0	0	140±7	2±0	116±3	496±35	377±65
Con A (2 µg/ml)	0±50	1200±230	3200±520	3220±410	272±19	88±8	1442±194	3973±419	31348±610
Con A (2 µg/ml) + MBA (5 mM)	0±100	80±50	0±600	200±700	228±4	32±11	187±14	649±81	1733±112

A Con A was added at staggered times, and the number of strand breaks in each sample was assayed simultaneously at 2h.

B MBA was added at t = 0 along with Con A. All assays were done at 48h. The activity of Na⁺ K⁺ ATPase was measured as incorporation of ⁸⁶Rb into suspended cells.

Table 2. Effect of MBA when added at various times after Con A stimulation on protein, RNA and DNA synthesis, and blast formation.

Time of addition of MBA (5 mM) after Con A stimulation	Incorporation of Macromolecular Precursors at 48h			
	³H Leucine	³H Uridine	³H Thymidine	Blasts
	cpm per 10⁶ cells			% of total
0h	187 ± 14	649 ± 81	1733 ± 112	32 ± 4
13h	275 ± 27	1498 ± 91	2617 ± 152	54.3 ± 4
26h	328 ± 99	1543 ± 124	2941 ± 44	60.6 ± 6
47.5h	1398 ± 65	3862 ± 112	30200 ± 943	86.7 ± 5

Con A was added to all cultures at t = 0.
MBA (5 mM) was added at various times after Con A.
All treatments were incubated until 48h then assayed with a 2h pulse of radioactive precursor.

produced during stimulation with mitogen and then rejoined
by a repair system regulated by the ADP ribose polymerase.
This would explain the strong inhibition produced by methoxybenzamide well after breaks were at a minimum.

This hypothesis was tested in an experiment in which
MBA was added to cultures of Con A-treated cells at progressively later times and DNA strand breaks assayed after
a standard incubation time (23h) (Table 3). An increasing
number of breaks was observed with increased time of incubation with MBA in both resting and stimulated cells.

These data (in Tables 1, 2 and 3) indicate that there
is a continuous production and repair of DNA strand breaks
in resting lymphocytes and that within 2h after mitogen
treatment there is a transient increase in ADP ribose
polymerase-mediated repair, which results in a decreased
steady state level of breaks in stimulated lymphocytes. An
alternate hypothesis is that MBA itself caused the DNA
strand breaks and that these accumulated with progressively
longer periods of treatment. This hypothesis is difficult
to retain since no additional breaks were induced by MBA in
cultures in which ADP ribosylation had already been 80-90%
inhibited by 3-amino-benzamide, or by heat shock (20 min at
45°C).

It was of interest to determine the factors that limit
rejoining of strand breaks in resting lymphocytes. One
possibility we examined was that NAD^+, the substrate for ADP
ribose polymerase, was limiting for ADP ribosylation and
thus for repair in resting lymphocytes.

This hypothesis was tested in a series of experiments
in which the intracellular concentration of NAD was increased 2-fold by incubating cells for 2h in medium containing its precursor, nicotinamide (300-400 μM) (Table 4).
Under these conditions many of the DNA strand breaks in
resting lymphocytes were rejoined in the absence of Con A.
This did not occur if MBA was added along with nicotinamide.
However, even though as many of the DNA strand breaks were
rejoined as in the case of the Con A-treated cultures, cells
did not manifest any other of the changes characteristic of
mitogenic activation, e.g. blast transformation or DNA synthesis. This indicates that repair of DNA strand breaks in
resting cells is necessary but not sufficient for lymphocyte
proliferation.

Table 3. Effect of MBA when added at various times after Con A stimulation, on
number of DNA strand breaks. Data shown represent the change in number
of strand breaks per diploid genome, compared to untreated control at
23h.

Time of addition of MBA (5 mM)	0h	6h	12h	18h	22h
Resting	3700 ± 500	1800 ± 500	1000 ± 450	260 ± 200	200 ± 300
Con A (2 μg/ml)	3600 ± 300	2000 ± 250	900 ± 200	300 ± 200	60 ± 200

Cell cultures with or without Con A were initiated at 6 = 0. MBA (5 mM) was
added at various times from 0 to 23h. All cultures were assayed for DNA strand
breaks at 23h.

Table 4. Induction of repair of DNA strand breaks in resting lymphocytes
by nicotinamide.

Treatment	Strand breaks repaired per diploid genome
Resting	0
Resting + 350 μM nicotinamide	2500±500
Resting + 350 μM nicotinamide + 5 mM MBA	0±180

All cultures were incubated at 37° for 2h then assayed for strand
breaks.

The hypothesis that NAD$^+$ may be limiting for repair in resting cells was supported by the finding that after incubation with Con A, the intracellular levels of NAD increased two-fold. Figure 1 shows that the two-fold increase in NAD coincided in time with an increase in ADP ribosylation measured in permeabilized cells. Both of these events coincided with the increase in repair induced by Con A.

DISCUSSION

Several recent reports have shown that resting lymphocytes contain DNA strand breaks (2,4). The fact that these breaks were rejoined following mitogenic stimulation was interpreted to mean that breaks produced during differentiation were definitively repaired allowing cells to cycle. However, it is unlikely that the simple presence of breaks in resting lymphocytes prevents their entry into the proliferative cycle because other agents such as 5-fluorouracil (100 μM) (1) or γ radiation (2,000 rads) (10) produced unrepaired DNA strand breaks; yet, in the presence of mitogen, treated cells were still able to undergo increased protein and RNA synthesis and blast transformation, although they then became blocked at the G_1/S boundary.

Our present data show that MBA can strongly inhibit lymphocyte proliferation even when added well after strand breaks are at a minimum; the fact that strand breaks accumulate with time of incubation with MBA in both resting and stimulated cells indicates that there is a dynamic equilibrium of breaking and rejoining of DNA. Therefore after mitogen treatment there is not simply a repair of DNA strand breaks but a change in equilibrium that results in a decreased steady state level of breaks. This decrease in number of breaks appears to result not from a decrease in production of breaks but from a transient increase in repair because it correlates in time with an increase in ADP ribosylase activity after mitogen treatment. Our data indicate that cellular levels of NAD are rate limiting for ADP ribosylation in resting lymphocytes and thus for repair of DNA strand breaks.

NOTE: Our conclusions concerning a continuous cycle of breaking and repair in lymphocytes are based on experiments in which ADP ribose polymerase was inhibited with methoxybenzamide. A recent report by Milam and Cleaver (11)

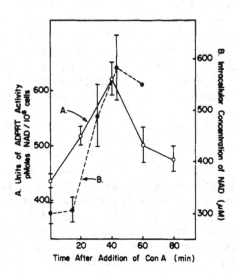

Fig. 1   Con A was added to cultures at staggered times so
that all assays were done simultaneously.

indicates that this compound has non-specific effects on
cell viability, glucose metabolism and DNA synthesis.
However, under the conditions used in our work, MBA produced only 4 to 10% cell death after 24h culture as measured by trypan blue exclusion, and, as shown in the text,
did not itself cause DNA strand breakage. The fact that
the drug had no effect on ^{3}H-thymidine incorporation when
it was added immediately prior to assay shows that it had
no effect on DNA synthesis. This, the fact that MBA caused
a parallel inhibition of ADP ribosylpolymerase activity, of
DNA repair and of blastogenesis and other data all indicate
that MBA prevented entry of cells into S phase by blocking
repair of endogenous DNA strand breaks.

REFERENCES

1. Boumah, C., Setterfield, G., and Kaplan, J.G. Can. J.
   Biochem. 62 (1984). In press.
2. Greer, W.J., and Kaplan, J.G. Biochem. Biophys. Res.
   Commun. 115, 834-840 (1983).
3. Farzeneh, F., Zalin, R., Brill, D., and Shall, S.
   Nature 300, 362-366 (1982).
4. Johnstone, A.P., and Williams, G.T. Nature 300, 362-
   366 (1982).
5. Williams, G.T., and Johnstone, A.P. Bioscience Reports
   3, 815-830 (1983).
6. Birnboim, H.C., and Jevcak, J.J. Cancer Research 41,
   1889-1892 (1981).
7. Ormerod, M.R. In Biology of Radiation Carcinogenesis
   (Ed. J.M. Yuhas, R.W. Tennant and J.D. Rogers), Raven
   Press, New York, pp. 67-90 (1976).
8. Berger, N.A., and Johnson, E.S. Biochim. Biophys. Acta
   125, 1-17 (1976).
9. Nisselbaum, J.S., and Green, S. Anal. Biochem. 27,
   212-217 (1969).
10. Roy, C., Brown, D.L., Lapp, W.A., and Kaplan, J.G.
    Immunology 163, 383 (1982).
11. Milam, K.M., and Cleaver, J.E. Science 223, 589-591
    (1984).

70 K DALTON PROTEIN SYNTHESIS AND HEAT SENSITIVITY OF

CHROMATIN STRUCTURE: DEPENDENCE ON CELL CYCLE

Roeland van Wijk[1] and Wiel Geilenkirchen[2]

[1]Department of Molecular Cell Biology and
[2]Department of Zoology, State University,
Padualaan 8, Utrecht, The Netherlands.

INTRODUCTION

Characteristic variations in nuclear morphology have been observed after growth stimulation (4) and during cell cycling (7). The nature, however, of these changes at the molecular level is largely unclear. The use of supranormal temperatures has been found to induce striking changes on the nuclear level (12). In this respect it is of interest to study the modifications of chromatin by hyperthermia at successive stages of the cell cycle. One approach has been to analyze the geometry and densitometry of nuclear images produced by differential staining of chromatin-DNA in situ by Feulgen reaction (3).

Here we report alterations of chromatin of Reuber H35 rat hepatoma cells both after serum stimulated growth and after hyperthermic treatment, and the differential effect of hyperthermia on chromatin of early G1 and G1/S cells. We already established in the Reuber H35 cell line that elevation of temperature increased the synthesis of a 70 kD protein (9). This paper extends these observations and describes the differential basal synthesis of this 70 kD protein after serum stimulation of growth. The results presented led to the hypothesis that the serum induced chromatin changes and the altered heat sensitivity of chromatin can be correlated with basal synthesis of 70 kD protein.

135

METHODS

Reuber H35 rat hepatoma cells were grown and synchronized (10). Heating of these cells and analysis of protein
synthesis, including preparation of 2-D-gels was as previously described (9). For quantitative spot analysis on
gels after 2-D-PAGE the spots were registered using a Quantimet 720 automated image analyzer. The Feulgen staining
procedure was modified (8). Feulgen-DNA measurements were
done with a Zeis scanning cytophotometer 01.

RESULTS

Proliferation of quiescent Reuber H35 cells. Serum
stimulated 80 percent of serum depleted quiescent cells,
stopped in G1. They performed one DNA cycle and division
synchronously. Cell cycle progression was determined auto-
radiographically (³H-thymidine) and by quantitative Feulgen
stain determination. Figure 1 is a composite figure showing
the incorporation of ³H-thymidine, the number of mitotic
figures and the frequency distribution of Feulgen staining
of H35 nuclei at various times after serum readdition. Continuous incorporation of thymidine was followed by autora-

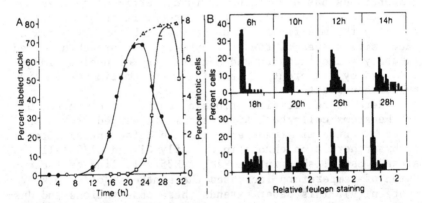

FIGURE 1. Stimulation of proliferation of resting H35 cells.
A. Percentage of labeled nuclei at various times following
serum addition and determined by 30 min pulse (●) or continuous (Δ) incorporation of ³H-thymidine. Mitotic cells
(□).
B. Percentage of cells with relative Feulgen staining at
various times following serum addition.
For each determination 60-100 nuclei were used.

diography and demonstrated a minimal lagperiod of 12 h before cells initiate DNA synthesis. During this lagperiod
cells have a G1 amount of DNA as was concluded from the
frequency distribution peak of Feulgen nuclear staining.
Incorporation of thymidine was found between 12 and 22 h.
The increased number of nuclei with twice the amount of
Feulgen staining confirmed the progression of cells through
their S and G2 phase. The culture showed G1 characteristics
again at 28 h after serum readdition.

   Feulgen stained nuclei: effect of hyperthermia. The
effect of incubation at 42ºC (30 min) on frequency distribution of Feulgen stained nuclei was studied at various
times after stimulation. Figure 2 shows the frequency distribution of Feulgen stained nuclei before and after heat
treatment. Cultures in early and mid G1 showed a broad frequency distribution after heat shock as compared to incubation at 37ºC. Because 30 min incubation of G1 cells at 42ºC
did not induce DNA synthesis, it is concluded that stainability for Feulgen has been increased after heat shock. In
G1 cultures at 28 and 34 h after serum stimulation a similar phenomenon was observed. In cultures at 12 till 20 h
following serum addition this broadening of the frequency
distribution after heat shock was diminished.

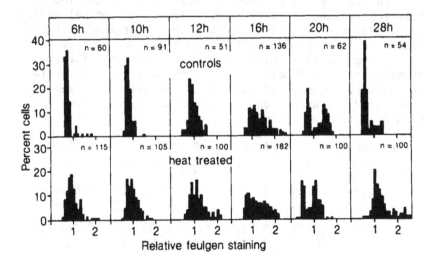

FIGURE 2. Percentage of cells with relative IOD after Feulgen staining at various times following serum addition.
Upper part: control cells at 37ºC. Lower part: Cells incubated 30 min at 42ºC prior fixation.

Geometric-densitometric image analysis. Geometric and
densitometric image analysis before and after heat treatment was done by measuring integrated optical density, pe-
rimeter, area, and computing variance and average optical
density (obtained by dividing the IOD by the area). Some
differences were found between the 6 and 12 h serum stimulated cultures. Table 1 shows the slight decrease of image
projection area in 12 h serum stimulated nuclei as compared
to 6 h (p = 0.05). In contrast AOD increased (p = 0.001).
The increased variance (p = 0.001) of 12 h nuclei means an
increasing occurrence of small nuclear areas with differential staining properties. This suggests a structural alteration of nuclei between 6 and 12 h after serum readdition.

When 6 and 12 h serum stimulated cultures have been
heat treated, considerable changes in nuclear geometry and
densitometry took place. Table 1 shows that 6 h nuclei show
50 percent reduction of area. The increased AOD is more
than twice the AOD of untreated cells and this is due to
the increased IOD after heat treatment as well as the area
decrease. The variance in 6 h nuclei has been increased
dramatically and illustrates the structural alteration of
the chromatin. Heat treatment of 12 h serum stimulated nuclei resulted in 33 percent decrease of area. The IOD of
heat treated nuclei is 16 percent less than control nuclei,
and explains the relatively low increase of AOD after heat
treatment (26 percent) as compared to 6 h nuclei. Furthermore, from the fact that the variance was not influenced
significantly it finally can be concluded that in contrast
to 6 h serum stimulated nuclei the untreated 12 h serum
stimulated cells show chromatin characteristics comparable
with heat treated chromatin.

Time	Temp.	Nuclear Area		AOD		Variance
h	$^\circ$C	μm^2	%	IOD/μm^2	%	
6	37	39.1±11.2	100	0.077+0.02	100	0.066+0.03
6	42	20.2+ 6.9	52	0.173+0.06	225	0.189+0.15
12	37	34.9+ 8.5	100	0.110+0.02	100	0.113+0.08
12	42	23.4+ 9.2	67	0.139+0.06	126	0.091+0.11*

TABLE 1. Geometric image analysis of nuclei and chromatin
in situ. The average optical density (AOD) = integreated
OD/Nuclear projection area, and the variance of OD per unit
area are given. *No significant difference between 37° and
42°C values.

The 70 kD protein synthesis. The data presented demonstrate that late G1-early S nuclei display an altered organization as compared to early or mid G1 cells and that hyperthermia hardly influences this structuring (although nuclear shape has been changed). This can be understood as expression of thermoresistant state. Recent data have indicated that heat tolerance is associated with heat shock proteins, among them hsp 70 is very prominent (9). Moreover, hsp 70 has been suggested to be able to act at the nuclear level (1). Therefore we considered it of interest to determine the synthesis of hsp 70 in quiescent Reuber H35 cells upon stimulation to proliferate.

The synthesis of hsp 70 was examined by two-dimensional fractionation of total cellular proteins labelled for 60 min with ^{35}S-methionine. The analysis of the electrophoresis gel image entails two steps: (1) the location of the hsp 70 protein, and (2) the measurement of the amount of protein present. Figure 3 demonstrates that hsp 70 fortunately is found in well defined location whose beginning and end is easy to define and not overlapped by any dark spot. In order to study basal, non-induced hsp 70 synthesis it is necessary to view longer exposures. Its presence was not detected in the autoradiograms of ^{35}S-labelled total cellular proteins of quiescent (Figure 3C) and 4 h serum stimulated H35 cells (not shown). In contrast at 12-13 h after serum stimulation 70 kD protein appears at the exact localization as heat induced 70 kD protein (Figure 3D). This protein was demonstrated in 19 h serum stimulated cultures (not shown) and was decreased slightly in relative intensity in 24 h cultures (Figure 3E).

Measurement of the amount of synthesized protein was done by measuring the optical density of the autoradiograph of a specific protein spot. It is then possible to compute its relative concentration as defined by its radioactivity. For this reason a radiation density standard was included which could be used for proteins in a range of 10^3-10^5 cpm depending on exposure time. Relative rates of synthesis were calculated by normalization which occurred by dividing the integrated intensity of the spot by the total of 200 spots. The relative rate of synthesis of this protein, measured at various times after serum readdition, is responsible for at least one pro mille of total protein synthesized at 12 and 19 h (Figure 3F).

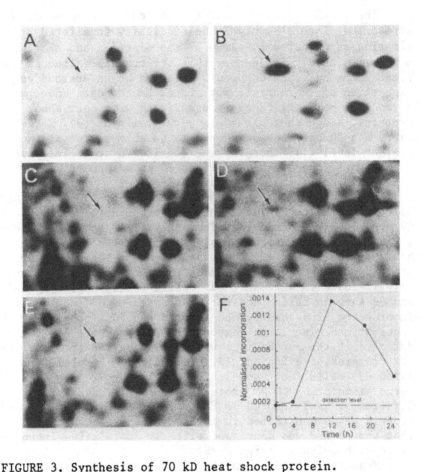

FIGURE 3. Synthesis of 70 kD heat shock protein.
A/B. Parallel cultures of randomly dividing H35 cells were incubated at 37°C with ^{35}S-methionine for 1 h (A), or at 42°C for 20 min, and subsequently reincubated in ^{35}S-methionine for 1 h at 37°C (B).
C/D/E. Synchronized H35 cells were incubated with ^{35}S-methionine for 1 h prior serum addition (C), or at 12 h (D) and 24 h (E) after serum addition. Each gel was prepared for fluorography and exposed to film for autoradiography. The arrows point to the localization of the hsp 70 spot.
F. Relative rate of synthesis of 70 kD protein (arrow-indicated in C, D and E). Rate of synthesis was determined by calculation of integrate density of autoradiographs after varying exposure times, and the use of a standardization curve. Rates are normalized as described in the text.

DISCUSSION

In this paper we have examined the nuclear morphology and intranuclear DNA organization of Reuber H35 hepatoma cells after growth stimulation by addition of serum in G1 cells and cells near the G1/S transition either with or without prior incubation at increased temperature. We observed a spatial redistribution during the progression of cells from early G1 to the G1/S transition. Recent data demonstrated that the changed Feulgen stained nuclei are characterized by the fact that the majority of the DNA is located in a narrow shell surrounding the nuclear and nucleolar borders (2). This redistribution resembles the chromatin restructuring of G1 cells following incubation at increased temperature. However, when cells in G1/S transition were heat treated no spatial redistribution was observed. The G1/S transition type of chromatin structure can be understood as a thermotolerant state. Thus, it shows some characteristics of a heat affected structure but had become relatively resistant towards a heat treatment.

The state of thermotolerance in general has been characterized now for protein synthesis inhibition, and heat induced cell death (9). A comparison of the proteins synthesized after heat shock with those synthesized by non-heated cells showed that the levels of synthesis of certain proteins were greatly enhanced following the heat treatment (6, 9). From these proteins the heat induced 70 kD protein has been described as a nuclear protein in Drosophila (1) and might be important for the thermotolerance of chromatin structure. But more intriguing were the recent findings that proteins of the same molecular weights are detected in normal cells (5, 11). Their data fit the model suggested by the results described in this paper showing that hsp 70 kD synthesis in normal cells was high during G1/S transition. It points to a molecular function of the 70 kD protein in the normal chromatin change after growth stimulation and its altered sensitivity for heat induced reorganization.

ACKNOWLEDGEMENTS

We wish to thank our colleagues Drs.A.A.M.S. van Dongen, A. van Meeteren, D.H.J. Schamhart for fruitful discussion and M. van Someren and C. Janssen-Dommerholt for carrying out part of the experimental work.

REFERENCES

1. Arrigo AP, Fakan S, Tissières A (1980) Localization of
   the heat shock-induced proteins in Drosophila melanogaster tissue culture cells. Dev Biol 78:86-103.
2. Belmont A, Kendall FM, Nicolini C (1984) Three-dimen-
   sional intranuclear DNA organization in situ: three
   states of condensation and their redistribution as a
   function of nuclear size near the G1-S border in HeLa
   S-3 cells. J Cell Sci 65:123-138.
3. Duijndam WAL, Duijn P van (1975) The influence of chromatin compactness on the stoichiometry of the Feulgen-
   Shiff procedure studied in model films. J Histochem
   Cytochem 23:891-900.
4. Kendall FM, Wu CT, Giaretti W, Nicolini CA (1977) Multiparameter geometric and densitometric analyses of the
   Go-G1 transition of WI-38 cells. J Histochem Cytochem
   25:724-729.
5. Kloetzel PM, Bautz EKF (1983) Heat shock proteins are
   associated with hnRNA in Drosophila melanogaster tissue
   culture cells. The EMBO Journal 2:705-710.
6. Li GC, Peterson NS, Mitchell HK (1982) Induced thermal
   tolerance and heat shock protein synthesis in Chinese
   hamster ovary cells. Int J Radiation Oncology Biol Phys
   8:63-67.
7. Linden WA, Fang SM, Zietz S, Nicolini C (1978) Chromatin study in situ. In Chromatin structure and function.
   Nicolini C editor, Plenum Co. 323-340.
8. Ploeg M van der, Duijn P van, Ploem JS (1974) High-re-
   solution scanning densitometry of photographic nega-
   tives of human chromosomes. Histochemistry 42:9-29
9. Schamhart DHJ, Walraven HS van, Wiegant FAC, Dongen
   AAMS van, Wijk R van (1982) Variations in some molecular events during the early phases of the Reuber H35
   hepatoma cell cycle. Biochimie 64:411-418.
10. Linnemans WAM, Rijn J van, Berg J van den, Wijk R van
    (1984) Thermotolerance in cultured hepatoma cells: cell
    viability, cell morphology, protein synthesis, and heat
    shock proteins. Radiat Res 98:82-95.
11. Velazquez JM, Sonoda S, Bugaisky G, Lindquist S (1983)
    Is the major Drosophila heat shock protein present in
    cells that have not been heat shocked ? J Cell Biol 96:
    286-290.
12. Wijk R van, Schamhart DHJ, Linnemans WAM, Tichonicky L,
    Kruh J (1982) Effect of hyperthermia on nuclear morpho-
    logy. Natl Cancer Inst Monogr 61:49-51.

INVOLVEMENT OF PROTEIN KINASES IN MITOTIC-SPECIFIC EVENTS

Margaret S. Halleck, Katherine Lumley-Sapanski,
Jon A. Reed and Robert A. Schlegel

Molecular and Cell Biology Program, The
Pennsylvania State University, University Park,
PA 16802

INTRODUCTION

Protein phosphorylation/dephosphorylation has been implicated in the mechanism of cytological events distinctive of mitosis, namely nuclear membrane breakdown and chromosome condensation. Three specific proteins of the nuclear lamina, called lamins, become highly phosphorylated attendent to nuclear dissolution, and then are dephosphorylated as the nuclear membrane reforms (1). Concomitantly, histones become highly phosphorylated as chromosomes condense (2).

Davis et al. (3) have recently presented evidence that mitotic-specific proteins recognized by monoclonal antibodies are phosphoproteins, and we have identified several protein kinase species which appear to be specific to mitotic cells (4). In this communication, we ask whether the behavior of these protein kinases in in vitro assays is that expected of the factors responsible for inducing mitotic-specific events based on the known characteristics of these so-called mitotic factors. In particular, we ask: a) Do the implicated protein kinases display the independence from regulation by cAMP expected? b) Does mixing of cellular extracts known to attenuate the biological activity of mitotic factors also attenuate protein kinase activity? c) Considering the autocatalytic nature of mitotic factors, can autophosphorylation of protein kinases be used to identify the molecular species responsible for inducing mitotic-specific events?

MATERIALS AND METHODS

Cytoplasmic extracts were prepared from populations of
mitotic or G1-phase HeLa cells as previously described (4).
Protein kinases were detected using an in situ gel assay
system (4) modified from previously published procedures (5-7).

SDS polyacrylamide gels (3% stacking, 10% running)
were prepared as in (4), with the inclusion of 0.1% SDS. In
some experiments an entire vertical lane was cut from a
non-denaturing gel, rehydrated in sample buffer containing
SDS and positioned horizontally across the top of an
SDS slab gel. In other experiments, horizontal bands were
cut from a rehydrated vertical lane of a non-denaturing gel
and each band positioned in a well of an SDS slab gel.
Electrophoresis was carried out at 50 V through the stacking gel, 200 V through the running gel. Gels were fixed,
stained, dried, then autoradiography performed to locate ^{32}P.

RESULTS

The system we employ to detect protein kinases
combines, sequentially, nondenaturing polyacrylamide gel
electrophoresis with an in situ phosphorylation assay using
histone as a recipient for $\gamma^{32}P$-ATP. As seen in Figure 1,
when autoradiography is performed on such a gel, bands of
protein kinase activity are found in the A, B and C regions
when extracts from mitotic cells are examined, in marked
contrast to extracts from interphase cells where protein
kinase activity is confined mainly to the C region, as
previously reported (4).

cAMP Dependency

A cAMP-independent protein kinase activity associated
with nuclear membrane preparations has been shown to selec-
tively phosphorylate a lamin-sized nuclear envelope polypeptide (8). Similarly, a protein kinase which phosphorylates histone H1 and whose activity is maximal at or near
mitosis has been shown to be cAMP-independent (9). We
have therefore asked whether the protein kinases we have
identified are cAMP dependent or independent. cAMP-
dependent protein kinases dissociate into their regulatory
and catalytic subunits in the presence of cAMP, thus acti-

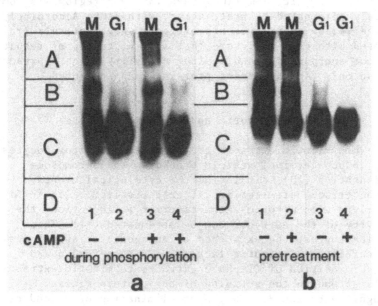

Figure 1. cAMP Dependency of Protein Kinases. Mitotic or
G1-phase extracts were separated on non-denaturing gels.
In (a), cAMP was absent (lanes 1 and 2) or present at 10 µM
(lanes 3 and 4) during in situ phosphorylation. In (b),
samples were incubated prior to electrophoresis without
(lanes 1 and 3) or with (lanes 2 and 4) 10 µM cAMP.

vating the catalytic subunit. We therefore carried out
phosphorylation reactions in the presence or absence of
10 µM cAMP and compared the extent of phosphorylation under
these conditions. As seen in Figure 1a, no differences in
the activities of mitotic extracts were readily apparent.
However, interphase extracts exhibited enhanced activity in
the B region in the presence of cAMP. Because the pI of
the catalytic subunit of cAMP-dependent protein kinases
precludes its entry into the gel, cAMP dependency can also
be determined by incubating extracts with cAMP prior to
electrophoresis and asking whether bands can still be
detected (7). As seen in Figure 1b, when mitotic extracts
were preincubated with 10 µM cAMP for 5 min at 37° C prior
to electrophoresis, conditions effective in preventing detection of commercial cAMP-dependent beef heart protein kinase in control gels, there was no change in the pattern of

bands observed. However, in extracts from interphase cells
the faint activity sometimes seen in the B region was completely eliminated by pretreatment with cAMP. Accordingly,
in the following experiments extracts were routinely preincubated with cAMP to insure that interpretation of results
was not confounded by this minor activity, which represented the only cAMP-dependent protein kinase detected.

Mixing of Mitotic and G1-Phase Extracts

When mitotic extracts are injected into frog oocytes,
they induce germinal vesicle breakdown and chromosome
condensation (10), simulating the cytological events
characteristic of mitosis. If extracts from cells in G1
are mixed with mitotic extracts prior to injection, the
activity of the extracts can be attenuated (11). These
results suggest the existence of antagonistic substances,
or inhibitors of mitotic factors. We therefore asked
whether addition of G1-phase extracts to mitotic extracts
would attenuate the activity of any protein kinase,
recognized by the diminution or elimination of a band of
activity.

Crude mitotic and G1-phase extracts were each adjusted
to equal protein concentrations, then mixed in different
proportions. Decreasing amounts of interphase extract were
added to a constant amount of mitotic extract such that the
former represented 50%, 33% and 20% of the total protein.
After incubation at room temperature for 15 min, the entire
contents of each mixture were loaded onto a gel. When
equal amounts of G1-phase and mitotic extracts were mixed,
the B region was obscured by overlap from the C region due
to the combined activities of both types of extracts (lane
2). However, when 1/2 or 1/4 as much interphase extract
was mixed with mitotic extract, the band in the B region
(less intense in the crude extract used in this experiment
than in the extract enriched by ammonium sulfate
fractionation used in Figure 1) could no longer be seen
(lanes 4 and 7), even though the same amount of mitotic
extract was applied to the gel as in lanes 3 and 6 where
mitotic extract alone was applied. In striking contrast,
all other bands in the mixtures appeared to be of the
intensities expected by a simple summation of the
activities of the G1-phase and mitotic extracts by
themselves (adjacent lanes on each side of mixtures).

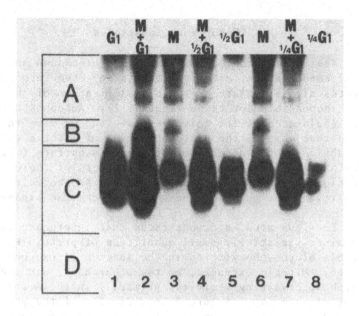

Figure 2. Effect of Mixing Mitotic and G1-Phase Extracts on Protein Kinase Activity. Crude mitotic and G1-phase extracts were each adjusted to 6.8 mg/ml of protein, mixed, incubated at room temperature for 15 min, followed by 5 min at 37° C in the presence of 10 µM cAMP, separated on nondenaturing gels and assayed for kinase activity. Lanes 3 and 6 contained 54.4 µg of mitotic extract; Lanes 1, 5 and 8 contained 54.4, 27.2 and 13.6 µg of G1-phase extract, respectively; Lanes 2, 4 and 7 contained 54.4 µg of mitotic extract plus 54.4, 27.2 or 13.6 µg of G1 phase extract, respectively, i.e., a mixture of the extracts in adjacent lanes.

This selective attenuation of the activity of a mitotic-specific, cAMP-independent protein kinase additionally nominates this species as a candidate for the biological activity exhibited by these extracts in oocyte assays. These experiments also point to the difficulty in preparing extracts with significant amounts of this activity since contamination by only a small fraction of G1-phase cells can so effectively prevent its detection. Turning these results to advantage, however, this system appears to provide a biochemical assay which may aid in the isolation of the inhibitory factor.

Endogenous Substrates

Besides inducing germinal vesicle breakdown and chromosome condensation, mitotic extracts injected into frog oocytes also stimulate an (auto)amplification of the factors responsible for these events (12). Autophosphorylation could account for this response, and we have in fact found that some of the protein kinases of mitotic extracts are active in the absence of exogenous substrate (4). An examination of the phosphorylated products of these reactions by SDS polyacrylamide gel electrophoresis might then reveal the polypeptides which compose the kinase.

Previous studies demonstrated that interphase extracts contained vanishingly small quantities of protein kinases capable of phosphorylation in the absence of exogenous substrate. Mitotic extracts, on the other hand, did present autophosphorylating activity, primarily in the A and B regions of gels (4). Figure 3 is a display of the phosphorylated products of such reactions, produced by cutting a lane from a nondenaturing gel and applying it to a second dimension SDS gel. Five polypeptides with approximate molecular weights of 105, 70, 58, 47 and 38kDa have been tentatively identified as originating from the B region of the non-denaturing gel, aligned at the top of the figure for reference. Only two of these proteins, 105 and 58kDa, appear to be unique to the B region, the other species being found in the A and C regions of the gel as well. Confirmation of these results was provided by cutting individual bands from non-denaturing gels and adding each to a separate well of an SDS gel (data not shown).

The phosphorylated polypeptides common to the A, B and C regions of the gel may perhaps represent endogenous substrates ubiquitous to all regions of the gel, or alternatively, subunits shared by each of the protein kinases. How the polypeptides unique to the mitotic-specific B region of the gel relate, if they do, to the mitotic-specific phosphoproteins of 182, 118, and 70 kDa identified by Davis et al.(3) or to the factor with an approximate molecular weight of 100 kDa responsible for the biological activity of mitotic extracts (11) is not clear and will require further investigations.

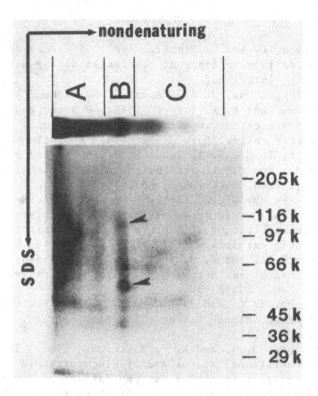

Figure 3: Molecular Weight of Proteins Phosphorylated in the Absence of Exogenous Substrate. A lane containing mitotic extract was cut from a non-denaturing gel, and the products phosphorylated in the absence of exogenous substrate were analyzed on a second dimension SDS gel. The excised lane has been mounted in its original position to identify the various regions of the non-denaturing gel. Arrows indicate polypeptides unique to the B region. Positions of molecular weight standards are indicated on the right- hand portion of the gel.

ACKNOWLEDGEMENTS

This research was supported by grant CD-111 from the American Cancer Society. R.A.S. is an Established Investigator of the American Heart Association. We thank Ramesh Adlakha for prompting us to undertake the mixing experiments.

REFERENCES

1. Gerace, L. and G. Blobel. 1980. The nuclear envelope
    lamina is reversibly depolymerized during mitosis.
    Cell. 19:277-287.

2. Gurley, L. R., J. A. D'Anna, S. S. Barham, L. L.
    Deaven, and R. A. Tobey. 1978. Histone phosphoryla-
    tion and chromatin structure during mitosis in Chinese
    hamster cells. Eur. J. Biochem. 84:1-15.

3. Davis, F. M., T. Y. Tsao, S. K. Fowler and P. N. Rao.
    1983. Monoclonal antibodies to mitotic cells. Proc.
    Natl. Acad. Sci. USA. 80:2926-2930.

4. Halleck, M. S., K. Lumley-Sapanski, J. A. Reed, A. P.
    Iyer, A. M. Mastro and R. A. Schlegel. 1984.
    Characterization of protein kinases in mitotic and
    meiotic extracts. FEBS Lett. 167;193-198.

5. Hirsch, A. and O. M. Rosen. 1974. An assay for
    protein kinase activity on polyacrylamide gels. Anal.
    Biochem. 60:389-394.

6. Knight, B. L. and J. F. Skala. 1977. Protein kinases
    in brown adipose tissue of developing rats. J. Biol.
    Chem. 252:5356-5362.

7. McClung, J. K. and R. F. Kletzien. 1981. The effect
    of nutritional status and of glucocorticoid treatment
    on the protein kinase isozyme pattern of liver paren-
    chymal cells. Biochim. et Biophys. Acta. 676:300-306.

8. Lam, K. S. and C. B. Kasper. 1979. Selective
    phosphorylation of a nuclear envelope polypeptide by an
    endogenous protein kinase. Biochem. 18:307-311.

9. Lake, R. S. and N. P. Salzman. 1972. Occurrence and
    properties of a chromatin-associated F1-histone phos-
    phokinase in mitotic Chinese hamster cells. Biochem.
    11:4817-4826.

10. Sunkara, P. S., D. A. Wright and P. N. Rao, 1979.
    Mitotic factors from mammalian cells induce germinal
    vesicle breakdown and chromosome condensation in
    amphibian oocytes. Proc. Natl. Acad. Sci. USA.
    76:2799-2802.

11. Adlakha, R. C., C. G. Sahasrabuddhe, D. A. Wright and
    P. N. Rao. 1983. Evidence for the presence of
    inhibitors of mitotic factors during the G1 period in
    mammalian cells. J. Cell Biol. 97:1707-1713.

12. Adlakha, R. C., C. G. Sahasrabuddhe, D. A. Wright, W.
    F. Lindsey and P. N. Rao. 1982. Localization of
    mitotic factors on metaphase chromosomes. J. Cell Sci.
    54:193-206.

PROTEIN KINASE ACTIVITY IN LYMPHOCYTES; EFFECTS OF

CONCANAVALIN A AND A PHORBOL ESTER

ANAND P. IYER, SHARON A. PISHAK, MARION J.
SNIEZEK, AND ANDREA M. MASTRO
THE PENNSYLVANIA STATE UNIVERSITY
MICROBIOLOGY PROGRAM
431 SOUTH FREAR LABORATORY
UNIVERSITY PARK, PA 16802

Lymphocytes treated with Concanavalin A (ConA) undergo a series of biochemical events leading to differentiation and division (reviewed by 1). Although the tumor promoter 12-0-tetradecanoylphorbol-13-acetate, TPA, itself is not mitogenic, it can enhance the response of bovine lymphocytes to lectins by acting as an analogue of interleukin 1 (2). Thus, in the presence of a suboptimal dose of ConA or a limiting number of macrophages, TPA allows full proliferation. However, exposing the lymphocytes to TPA for about 24 hrs before stimulation with ConA greatly depresses the response (3). Furthermore, TPA blocks the mixed lymphocyte response (4).

The biochemical changes which occur after ConA stimulation have been reported to include changes in protein kinase activity and in protein phosphorylation (5,6,7). Furthermore, the cell receptor for TPA is reported to be a protein kinase C (8,9). TPA can activate protein kinase C in vitro (10). Thus, based on the possible importance of phosphorylation in the regulation of DNA synthesis in lymphocytes treated with ConA or TPA, we examined the protein kinase activity. In order to screen for several protein kinases at once even with small sample sizes, we assayed activity in situ in nondenaturing polyacrylamide gels. We modified existing techniques (11,12,13) to produce a high resolution system to visualize both substrate dependent and independent protein kinase activity (14).

MATERIALS AND METHODS

Lymphocytes isolated from bovine lymph nodes were cultured at 1 x 10^7 cells/ml (3), plus or minus ConA (7.4 μg/
ml) or TPA (10^{-7}M). Twenty-four hours later the cells (50-
150 ml) were collected on ice, centrifuged from the medium
(300 x g for 10 min) and washed three times in phosphate-
buffered saline (PBS) before being resuspended and sonicated
in 2.0 ml of a buffer of 50 mM Tris-HCl, pH 7.4; 50 mM β-
mercaptoethanol; 2 mM EGTA; 2 mM EDTA; 1 mM phenylmethyl-
sulfonyl fluoride (PMSF); 10 mM NaF; and 0.1% Triton X-100.
The sonicated samples were held on ice for 30 min with occa-
sional shaking before being centrifuged at 100,000 x g for
1 h (Beckman L565, SW 50.1 rotor). Supernatants were collected, aliquoted and frozen at -80°C. Before electrophoresis, samples were dialyzed against 0.1 M Tris-Cl (pH 6.8),
2 mM PMSF.

Electrophoresis and phosphorylation were carried out as
previously described (14). In brief a modified Laemmli gel
system (15) without SDS was used. A 5% acrylamide running
gel (9 cm length) containing 10% glycerol, and a 3% acrylamide stacking gel with 6% glycerol (1 cm) were polymerized
onto a gel support film (Gel Bond, FMC Corp.). Samples,
100 μg protein, were electrophoresed at 50V through the
stacking gel and 200V through the running gel in the cold
for approximately 4 1/2 h. Following electrophoresis the
gels were incubated for 1 h each in two changes of 50 ml
ice-cold 50 mM Tris-HCl (pH 7.5). For phosphorylation a
slab gel or section of a gel was incubated for 30 min at
37°C in 25 ml of a solution of 0.3 M Tris-acetate (pH 7.4),
0.04 M Mg-acetate, 0.15 M NaF, 1 mM sodium phosphate buffer
(pH 7.4), 0.3 mM EGTA (16), plus or minus histone type II-AS
(Sigma) at 4 mg/ml. The reaction was initiated by the addition of [γ^{32}P]ATP, 2-3 Ci/mM, 2 μCi/ml. After an incubation
with gentle shaking at 37°C for 1 h, the reaction was terminated by removal of the incubation mixture followed by two
rapid rinses with 100 ml of cold 5% trichloroacetic acid, 1%
phosphoric acid. The gel was soaked overnight in the acid
solution and washed three more times with the same solution
before it was fixed overnight in 10% acetic acid, 30% methanol, and dried in an oven at 60°C. Dried gels were exposed
to Kodak X-omat R film for radioautography in a cassette
containing one detail screen (Cronex XTRA Life Detail,
DuPont), for about 6-24 h, depending on the radioactivity.
The radioautograph was scanned on an LKB Laser Densitometer
(Model 2202 Ultroscan).

RESULTS AND DISCUSSION

Six bands of protein kinase activity were visible when
cytoplasmic extracts from ConA-stimulated lymphocytes were
separated on non-denaturing gels and assayed in situ using
histone as a substrate (Figs. 1,2). Bands 2,3,4 and 5 were
the most intense. Band 3 appeared to be a doublet. Bands
1 and 6 were barely detectable in short exposures of radio-
autographs but became visible on longer exposures. Although
these bands were present in stimulated and nonstimulated
cells, they were of greater intensity in the stimulated
cells. Additional kinase activity evident in band 2, was
present in the ConA-treated cells, but faint or absent in
the nonstimulated. Thus, the appearance of band 2, as well
as an increase in intensity of the other bands, was most
characteristic of the stimulated cells, using histone as
the exogenous substrate.

When homogenates of lymphocytes treated for 24 h with
TPA were assayed, the pattern of phosphorylation in the presence of histone was generally the same as that of the unstimulated cells (Fig. 1,2). One exception was a decrease
in intensity of band 4. When histone was omitted from the
phosphorylation mix for any of the samples, two faint bands
were present (Fig. 1). These autophosphorylating proteins
migrated in the same positions as bands 3 and 4, visible in
the presence of histone.

In order to identify cAMP-dependent kinases, the extracts were incubated with 10^{-4}M cAMP before electrophoreses.
This treatment separates the catalytic and regulatory subunits of cAMP-dependent protein kinase and prevents the catalytic subunit from migrating into the gel (11,12). Thus,
cAMP dependence is evidenced by the loss of a band(s) of
activity. After treatment of the extracts of lymphocytes
stimulated with ConA with cAMP, three bands of activity
(bands 2,5,6) disappeared from the gels (Fig. 1). Bands 3
and 4 did not disappear after treatment with cAMP (Fig. 2)
or cGMP (data not shown). These kinases appear to belong to
the general class of cAMP-independent kinases.

In order to determine the actual number of proteins in
an area of kinase activity, each band was cut from the native
gel after phosphorylation in the presence or absence of histone, rerun on an SDS gel, and exposed to x-ray film. As
expected, the results for all six bands were similar; the

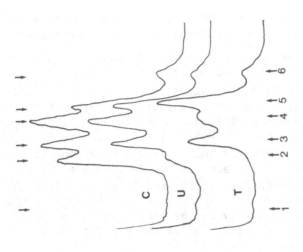

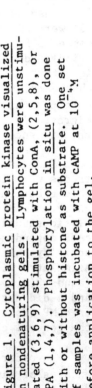

Figure 1. Cytoplasmic protein kinase visualized on nondenaturing gels. Lymphocytes were unstimulated (3,6,9) stimulated with ConA, (2,5,8), or TPA (1,4,7). Phosphorylation in situ was done with or without histone as substrate. One set of samples was incubated with cAMP at 10^{-4}M before application to the gel.

Figure 2. Tracing of a radioautogram of a gel showing protein kinase activity of lymphocytes in the presence of histone. C, ConA; U, untreated; T, TPA.

major bands visible by silver staining and by radioautography were the histones, the added endogenous substrate
(data not shown). In addition there were seven or eight
other faintly labeled bands, some of which are contaminants of the histone preparation. The rest may be endogenous autophosphorylating components or substrates in the
cell lysate.

In order to look at the autophosphorylating components
more closely, the same procedure was carried out with band
3 from a gel phosphorylated without histone. The radioautogram revealed two labeled bands estimated to be approximately 78 and 58 KD (Fig. 3). Interestingly, protein kinase
C is reported to be an autophosphorylating kinase of about
75-80 KD with a smaller subunit of about 55 KD (17). The
possibility that band 3 is a protein kinase C needs to be
determined.

In order to examine the endogenous substrates for the
cytoplasmic kinases, the extracts were phosphorylated in a
test tube without added substrates or kinases. After separation on an SDS gel, numerous phosphorylated substrates
were seen by radioautography (Fig. 4). The ConA-stimulated
cell extract showed at least 25 bands. cAMP increased the
intensity of several of them (approximately 83,80,61,57,50,
49 to 39,29 and 24 KD). One band, approximately 80 KD, was
one of the most intense in the stimulated cells and may
correspond to an 80 KD protein which becomes highly phosphorylated in growth-factor stimulated 3T3 cells (18). The
unstimulated lymphocytes showed a pattern similar to that
of the stimulated cells but the intensity of the bands was
much lower. However, several bands which were enhanced by
cAMP in the ConA-stimulated cells were absent or barely
detectable in the unstimulated lymphocytes (approximately
57,44 to 50,39 and 24 KD). It is not known at this time
if the protein kinase of band 2 seen in the nondenaturing
gel is responsible for the phosphorylation of these substrates. Addition of TPA directly to the phosphorylation
mixture had no effect on the phosphorylation.

In summary, we have used a system to assay endogenous
kinases in lymphocytes. Although there is evidence that
lymphocytes have all of the machinery for regulation by
phosphorylation of proteins with protein kinases, there has
been no clear resolution of the numbers or role of either
(see ref. 1 for review). Standard DEAE-chromatography

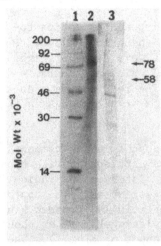

Figure 3. SDS gel of band 3 kinase from ConA-stimulated lymphocytes after phosphorylation on a nondenaturing gel. Band 3 was cut from a nondenaturing gel after phosphorylation in the absence of histone, and rerun on a 10% acrylamide SDS gel. MW markers were ^{14}C-methylated myosin, phosphorylase B, bovine serum albumin, ovalbumin, carbonic anhydrase and lysozyme. Lane 1, markers; Lane 2, radioautograph; Lane 3; silver stain.

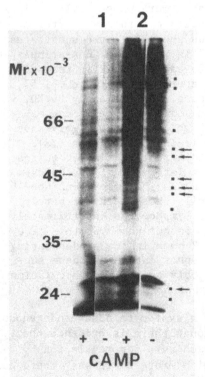

Figure 4. Radioautograph of endogenous substrates in cytoplasmic extracts of unstimulated and ConA-stimulated lymphocytes. The cytoplasmic fractions were phosphorylated in vitro (16) and separated on SDS gels. When present cAMP was 10^{-6}M. 1, unstimulated; 2, ConA. (■) cAMP dependent; (←) present only in ConA stimulated.

techniques are not only difficult to use with small samples, but also have not given good resolution of some classes of kinases in extracts of various cultured cells (11,12) or with lymphocytes (5). Although several techniques for assaying protein kinases on gels have been published (11, 12,19,20), we have modified the technique to increase its sensitivity and resolution. We have detected two autophosphorylating and six histone-dependent kinases. At least three bands of kinase activity were sensitive to cAMP. One of these, was intensified in ConA-stimulated cells. Two major bands were not cAMP-dependent. One of these may be protein kinase C, a cAMP-independent kinase and the most abundant kinase in lymphocytes (21). Another cAMP independent band was depressed when cells were treated with TPA under the conditions where proliferation was also depressed. Kwong and Mueller (22) have also reported loss of phosphorylation of a membrane protein after treatment of lymphocytes with TPA, but we do not know how this change relates to the decrease in band 4 activity. Further correlation between the kinases and the stimulatory and inhibitory reactions remains to be defined.

ACKNOWLEDGMENTS

The work was supported by a grant, CA-24385, NCI, U.S. Department HHS. A.M.M. is the recipient of an RCDA, CA-00705, from the same source.

REFERENCES

1. Hume, D.A. and M.J. Weidemann, (1980)
    Elsevier/North-Holland Biomedical, New York.
2. Mastro, A.M. and K.G. Pepin, (1982) Cancer Res.
    42:1630-1635.
3. Mastro, A.M. and K.G. Pepin (1980) Cancer Res.
    40:3307-3312.
4. Mastro, A.M., T. Krupa and P. Smith (1979) Cancer Res.
    39:4078-4082.
5. Chaplin, D.D., H.J. Wedner and C.W. Parker (1979)
    Biochem. J. 182:525-537.
6. Chaplin, D.D., H.J. Wedner and C.W. Parker (1980) J.
    Immunol. 124:2390-2398.
7. Wang, T., J. Foker, and A. Malkinson (1981) Exp. Cell
    Res. 134:409-415.
8. Ashendel, C.L., J.M. Staller, and R.K. Boutwell (1983)
    Cancer Res. 43:4333-4337.
9. Niedel, J.E., L.J. Kuhn and G.R. Vandenbark (1983)
    Proc. Nat. Acad. Sci. 80:36-40.
10. Castagna, M., Y. Takai, K. Kaibuchi, K. Sano, U.
    Kikkawa and Y. Nishizuka (1982) J. Biol. Chem.
    257:7847-7851.
11. Knight, B.L. and J.P. Skala (1977) J. Biol. Chem.
    252:5356-5362.
12. McClung, J.K. and R.F. Kletzien (1981) Biochim.
    Biophys. Acta 676:300-306.
13. McClung, J.K. and R.F. Kletzien (1981) Biochim.
    Biophys. Acta 678:106-114.
14. Iyer, A.P., S.A. Pishak, M.J. Sniezek and A.M. Mastro
    (1984) Biochem. Biophys. Res. Comm. (in press).
15. Laemmli, U.K. (1970) Nature 227:680-685.
16. Mastro, A.M. and E. Rozengurt (1976) J. Cell Biol.
    251:7899-7906.
17. Kikkawa, U., R. Minakuchi, Y. Takai and Y. Nishizuka
    (1983) Meth. Enzymol. 99:288-298.
18. Rozengurt, E., M. Rodriguez-Pena, and K.A. Smith
    (1983) Proc. Nat. Acad. Sci. 80:7244-7248.
19. Hirsch, A. and O.M. Rosen (1974) Anal. Biochem.
    60:389-394.
20. Gagelmann, M., W. Pyerin, D. Kübler and V. Kinzel
    (1979) Anal. Biochem. 93:52-59.
21. Ogawa, Y., Y. Takai, Y. Kawahara, S. Kimura and Y.
    Nishizuka (1981) J. Immunol. 127:1369-1374.
22. Kwong, C.H. and G.C. Mueller (1983) Carcinogenesis
    4:663-670.

EFFECT OF GOSSYPOL ON DNA SYNTHESIS

IN MAMMALIAN CELLS

Potu N. Rao, Larry Rosenberg and

Ramesh C. Adlakha

Department of Chemotherapy Research, The

University of Texas M. D. Anderson Hospital and

Tumor Institute, Houston, Texas 77030.

Gossypol, a yellow phenolic compound, extracted from cotton seed and plant parts has been shown to be an effective male contraceptive in China, which induces structural abnormalities in sperm and reduces the sperm count by 99.9% in the subjects tested (8). Eventhough gossypol does not appear to be a mutagen according to the Ames test (3) and the sperm head abnormality assay in mice (6), it has been shown to reduce the mitotic index in phytohemagglutinin-stimulated human peripheral blood lymphocytes (14). Gossypol is known to inhibit the activity of several enzymes, including various dehydrogenases, ATPase, and other enzymes involved in mitochondrial oxidative phosphorylation (1,5,7,9). Gossypol was reported to have no effect on the incidence of chromosome breakage or ploidy but reduced the mitotic index and the rates of DNA, RNA and protein synthesis in Chinese hamster ovary (CHO) cells and in human lymphocytes (16). Subsequently Wang and Rao (15) have shown that gossypol is a specific inhibitor of DNA synthesis in cultured cells. In the prescence of the drug, cells can enter S phase but fail to complete replication. Hence the objective of the present study was to examine the mechanism for the gossypol-induced inhibition of DNA synthesis in mammalian cells.

EFFECT OF GOSSYPOL ON CELL CYCLE PROGRESSION

Wang and Rao (15) made a detailed study of the effects
of gossypol on cell cycle progression and DNA synthesis in
CHO cells and HeLa cells in culture. They observed that
the continued presence of gossypol (10 µg/ml) in the growth
medium did not inhibit the progression of G1 cells into S
phase or G2 cells into mitosis. Most of those in S phase,
and the G1 cells that entered S phase subsequent to the addition of the drug failed to reach mitosis. Only a small
proportion of the cells in S phase, entered mitosis, probably those in late S phase, at the time of addition of the
drug. These results were further confirmed by using HeLa
cells synchronized in various phases of the cell cycle.
Gossypol had no effect on the progression of synchronized
mitotic cells into G1, G1 cells into S phase (Fig. 1A), and
G2 cells into mitosis (Fig. 1C). When gossypol was added
to cells in early S phase, i.e., 1 hr after reversal of the
second thymidine block, there was a dose-dependent inhibition of their progression into mitosis (Fig. 1B). The
mitotic accumulation in the presence of Colcemid was about
20% at a drug concentration of 10 µg/ml and 6% at the 20 µg/
ml dose. To determine the point in the cell cycle where
the cells are arrested, gossypol-treated CHO cells (10 µg/
ml for 48 h) were fused with mitotic cells and the pre-
maturely condensed chromosomes (PCC) were classified as
G1-, S-, or G2-PCC based on their morphology as described
earlier (13). Almost all of the cells in gossypol-treated
cultures were arrested in S phase as indicated by their
characteristic pulverized appearance. However, very few of
them incorporated ^{3}H-thymidine as revealed by autoradiography when pulse labeled with ^{3}H-TdR for 30 min before
fusion. These data indicate that in the presence of gossypol cells can enter S phase, but most of them are unable
to complete replication and proceed to mitosis (15).

EFFECT OF GOSSYPOL ON DNA SYNTHESIS

Wang and Rao (15) have also shown that gossypol at a
concentration of 10 µg/ml had no effect on the rates of
synthesis of RNA and proteins, but DNA synthesis decreased
to approximately 50% and 25% of the control by 1 h and 8 h,
respectively. DNA synthesis was reduced to zero by 24 h of
treatment (Fig. 2).

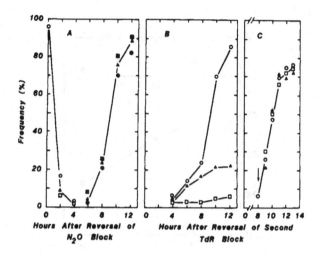

Fig. 1. Effect of gossypol on the cell cycle traverse of synchronized HeLa cells. O,△,□,mitotic index; ●,▲,■, labeling index. Cells were continuously exposed to gossypol (10 µg/ml) (△), and gossypol (20 µg/ml)(□),throughout the course of the experiment. O , control. A, effect on the progression of synchronized mitotic cells into S phase. [3H]-thymidine (0.1 µCi/ml) and various concentrations (0, 10 and 20 µg/ml) of gossypol were added to the dishes at t = 0 hr. Colcemid (0.05 µg/ml) was added at 4 hr after reversal of N2O block. B, effect on the progression of synchronized S cells into mitosis. Colcemid (0.05 µg/ml) and various concentrations of gossypol were added at t = 0 hr. C, effect on the progression of G2 cells into mitosis. Colcemid (0.05 µg/ml) and various concentrations of gossypol were added at 8 hr after the reversal of the second thymidine block (arrow). (From Ref. 15).

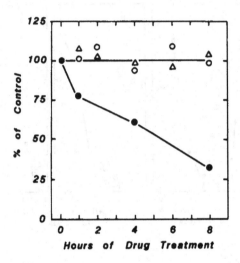

Fig. 2.   Effect of gossypol on DNA, RNA and protein synthesis in HeLa cells. Two sets of 18 each of 35-mm dishes
were plated with equal numbers of cells (approximately,
1 x 10^6 cells/dish) 1 day before the experiment. The experiment was begun by adding gossypol (10 μg/ml) to one
set of dishes, while the other set served as control. Colcemid (0.05 μg/ml) was added to both sets of dishes to
block cell division and to hold the cell number constant.
At various times after the addition of gossypol, dishes
from the control and drug-treated cultures were pulse
labeled for 30 min with [^{3}H]-thymidine (1.0 μCi/ml) and
processed for scintillation counting to measure the incorporation of label into DNA (●), RNA (○), and protein
(△), respectively. (From Ref. 15).

THE POSSIBLE TARGET OF GOSSYPOL'S ACTION

In order to ascertain if gossypol interacts directly with DNA thus rendering it inaccessible or unsuitable for replication, the following experiment was performed. Unlabeled HeLa cells (U) irreversibly blocked in S phase by a prolonged gossypol treatment (10 µg/ml for 48 h) were fused with ^{3}H-TdR-labeled HeLa cells synchronized in S phase (L) by excess thymidine double block method (11). Following cell fusion using UV-inactivated Sendai virus (12),the cells were resuspended in culture medium containing Colcemid (0.1 µg/ml) and ^{3}H-TdR (0.1 µCi/ml; sp. act. 6.7 Ci/mmole). Cell samples were taken at regular intervals, fixed processed for autoradiography, stained, and scored for the frequency of various types of binucleate cells (i.e., L/L, both nuclei labeled; U/U, both nuclei unlabeled; and L/U, one nucleus labeled and the other unlabeled). We observed a rapid decrease in the frequency of the L/U class, i.e., those formed by the fusion of gossypol-arrested cells with synchronized S phase HeLa cells, with a concomitant increase in the L/L class of binucleate cells. These results indicate that DNA synthesis can be reinitiated in gossypol-arrested cells following fusion with S phase cells. These results also suggest that gossypol may not be acting on DNA per se but rather may affect the inducers of DNA synthesis or some of the various enzymes involved in DNA replication.

EFFECT OF GOSSYPOL ON DNA POLYMERASE α

Since DNA polymerase α is an important enzyme involved in the replication of DNA, we studied the effect of gossypol on the activity of this enzyme. DNA polymerase activity was assayed using the purified enzyme of Micrococcus luteus (Sigma) or whole cell extracts made from 20 x 10^6 HeLa cells according to Pendergrass et al (10). Cell extracts were stored at -70°C in small aliquots and thawed as needed. The assay used was similar to that of Grosse and Krauss (4). The enzyme was added to a reaction mixture containing 50 mM potassium phosphate buffer (pH 7.1), 10 mM $MgCl_2$, 2 mM dithiothreitol, 102 µM each of dATP, dGTP and dTTP, 22 µM dCTP, 20 pmole ^{32}P-dCTP (sp. act. 3000 Ci/mmole) and 200 µg/ml activated calf thymus DNA (2) in a total volume of 60 µl and incubated at 37°C for 1 h. Where indicated, various concentrations of gossypol or metal salts were included at the start of the assay. Samples of 10 µl

were removed at intervals and spotted onto Whatman filter
paper discs pretreated with 0.1 M pyrophosphate. The discs
were immediately immersed in cold 5% trichloroacetic acid
(TCA) to stop the reaction, then batch-washed 3 times in
5% TCA (10 ml per disc) and twice in 95% ethanol before
air drying. TCA-precipitable counts were measured in a
Packard scintillation counter (Model 2650) using 10 ml
of scintillation fluid per disc. Greater than 85% of enzyme activity measured by this assay in cell extracts was
due to DNA polymerase α, as determined by inhibitor studies
using aphidicolin and dideoxy TTP and by sucrose gradient
sedimentation (data not shown).

Gossypol had an immediate inhibitory effect on the
activity of the purified DNA polymerase of M. luteus reducing it to <2% of control at a concentration of 5 µg/ml
(Table 1). In contrast, the enzyme activity in the crude
extracts of HeLa cells was relatively less sensitive to
inhibition by gossypol. A much higher concentration of
gossypol was required to cause a significant reduction in
the activity of DNA polymerase α. This is probably due to
the presence of other proteins or factors in the HeLa cell
extracts that might bind to gossypol and render it less
active.

To study the kinetics of inactivation of DNA polymerase α by gossypol, random populations of HeLa cells were
exposed to gossypol (10 µg/ml) for various lengths of time,
cell extracts prepared, and assayed for enzyme activity as
described earlier. The activity of this enzyme in the extracts remained relatively high (74% of control) even after
incubation of cells with gossypol for 8 h. A 16 h incubation of cells with gossypol reduced the activity of DNA
polymerase α to 31% of control. These results indicate
that DNA polymerase α is inhibited by gossypol in a time-
dependent manner. It is also possible that gossypol may
have an effect on enzymes other than DNA polymerase α that
are involved in DNA synthesis.

REVERSAL OF THE GOSSYPOL-INDUCED INHIBITION OF DNA SYNTHE-
SIS IN HeLa CELLS BY Fe^{2+}

Since catechols are known to chelate metals and as
gossypol has a double catechol structure, we decided to examine whether the effects of gossypol on DNA synthesis are
due to its chelating action. HeLa cells were first grown

TABLE 1

EFFECT OF GOSSYPOL ON THE ACTIVITY OF DNA POLYMERASE

Treatment	Enzyme activity[a]		% of control	
	M.luteus	HeLa	M.luteus	HeLa
Control	78	27	100	100
Gossypol 5 µg/ml	0.6	22	0.8	81
Gossypol 10 µg/ml	0.5	16	0.6	59
Gossypol 20 µg/ml	1.0	4	1.3	15
Gossypol 50 µg/ml	0	0	0	0
DMSO	88	–	102	–

[a]Activity is expressed as p mole dCTP incorporated into activated calf thymus DNA per hour at 37°C.

in the presence of different concentrations (20 µM–5 mM) of the various metal salts, to determine the concentrations at which these metals have no cytotoxic effects on cells. An equal number (5 x 10^5) of exponentially growing HeLa cells were plated in a number of 60 mm dishes a day before the experiment. Gossypol (5 µg/ml) and/or different concentrations (50 µM–1 mM) of $FeSO_4$, $ZnSO_4$, $CuSO_4$ or $MgCl_2$ were separately added to the HeLa cells. Untreated dishes served as controls. Colcemid (0.05 µg/ml) was added to all the dishes to keep the cell number constant. After incubation for 8 h cells were labeled with ^{3}H-thymidine (1 µCi/ml) for 1 h. Cells were processed for scintillation counting to measure the uptake of TCA-precipitable ^{3}H-thymidine into HeLa cells as described earlier (15). Of all the metals tested only Fe^{2+} could reverse the gossypol-induced inhibition of DNA synthesis (Table 2). These results suggest that gossypol may be blocking DNA synthesis by chelating Fe^{2+}, thus rendering it unavailable for the normal function of enzymes involved directly or indirectly in DNA replication.

TABLE 2

REVERSAL OF GOSSYPOL-INDUCED INHIBITION OF
DNA SYNTHESIS BY FeSO$_4$

Treatment	TCA-precipitable cpm (% of control)
A random culture of HeLa cells (control)	100.00
Control + gossypol (5 µg/ml)	14.86
Control + FeSO$_4$ (100 µM)	93.08
Control + gossypol (5 µg/ml)+FeSO$_4$ (100 µM)	92.48
Control + FeSO$_4$ (500 µM)	98.51
Control + gossypol (5 µg/ml) + FeSO$_4$(500 µM)	109.03
Control + FeSO$_4$ (1 mM)	106.20
Control + gossypol (5 µg/ml) + FeSO$_4$(1 mM)	99.40
Control + MgCl$_2$ (500 µM)	109.48
Control + gossypol (5 µg/ml) + MgCl$_2$ (500 µM)	14.26
Control + ZnSO$_4$ (200 µM)	83.15
Control + gossypol (5 µg/ml) + ZnSO$_4$ (200 µM)	10.74
Control + CuSO$_4$ (500 µM)	111.79
Control + gossypol (5 µg/ml) + CuSO$_4$ (500 µM)	16.21

HeLa cells were treated with gossypol in the presence or
absence of different metal salts for 8 h, pulse labeled
with ^{3}H-thymidine(1 µCi/ml) for 1 h and processed for
scintillation counting.

In summary, we have shown that gossypol is a potent
inhibitor of DNA synthesis, but does not affect RNA or protein synthesis at the doses tested. In the presence of the
drug cells are specifically blocked in S phase, whereas
cells in other phases of the cell cycle are unaffected.
Cell fusion experiments suggest that DNA per se is not
directly affected since the inhibition of DNA synthesis can
be reversed. Gossypol inhibited DNA polymerase from M.
luteus and HeLa cells in a dose-dependent manner. The
gossypol-induced inhibition of DNA synthesis in HeLa cells
could be reversed by the addition of FeSO4 to the media.
These results suggest that inhibition of DNA polymerase by

gossypol may account in part for the inhibition of DNA
synthesis observed in HeLa cells. However, inhibition by
gossypol of other enzymes involved in DNA synthesis is not
ruled out.

ACKNOWLEDGEMENTS

We thank Josephine Neicheril for her excellent
secretarial assistance in preparing this manuscript, Dr.
N. Burr Furlong for development of the DNA polymerase
assay, and Dr. Walter N. Hittelman for encouragement and
support. Supported in part by research grants CA-27544
from the National Cancer Institute, ACS-CH-205 from the
American Cancer Society, and a Rosalie B. Hite Fellowship
to L.J.R.

REFERENCES

1. Abdou-Dania, M.B., and J.W. Dieckert. 1974. Gossypol:
   uncoupling of respiratory chain and oxidativephospho-
   rylation. Life Sciences 14:1955-1963.
2. Aposhiar, H.V., and A. Kornberg. 1962. The polymerase
   formed after T2 bacteriophage infection of E. Coli:a
   new enzyme. J. Biol. Chem. 237:519-525.
3. Colman, N., A. Gardner, and V. Herbert. 1979. Non-
   mutagenicity of gossypol in the Salmonella/mammalian-
   microzome plate assay. Environ. Mutage. 1:315-320.
4. Grosse, F., and G. Krauss. 1981. Purification of a 9S
   DNA polymerase α species from calf thymus. Biochem.
   20:5470-5475.
5. Kalla, N.R., and M. Vasudev. 1981. Studies on the male
   antifertility agent-gossypol acetic acid. II. Effect
   of gossypol acetic acid on the motility and ATPase
   activity of human spermatozoa. Androlog. 13:95-98.
6. Majumdar, S.K., H.J. Ingram, and D.A. Prymowicz. 1982.
   Gossypol - an effective male contraceptive was not
   mutagenic in sperm head abnormality assay in mice. Can.
   J. Genet. Cytol. 24:777-780.
7. Montamat, E.E., C. Burgos, N.M. de Burgos, L.E. Rovai,
   A. Blanco, and E.L. Segura. 1982. Inhibitory action
   of gossypol on enzymes and growth of Trypansoma cruzi.
   Science 218:288-289.
8. National Coordinating Group on Male Infertility Agents
   (China). 1979. Gossypol - a new antifertility agent
   for males. Gynecol. Obstet. Invest. 10:163-176.

9. Olgiati, K.L., and W.A. Toscano, Jr. 1983. Kinetics
    of gossypol inhibition of bovine lactate dehydrogenase
    X. Biochem. Biophys. Res. Commun. 115:180-185.
10. Pendergrass, W.R., A.C. Saulewicz, G.C. Burmer, P.S.
    Rabinovitch, T.H. Norwood, and G.M. Martin. 1982.
    Evidence that a critical threshold of DNA polymerase
    α activity may be required for the initiation of DNA
    synthesis in mammalian cell heterokaryons. J. Cell.
    Physiol. 113:141-151.
11. Rao, P.N., and J. Engelberg. 1966. Effects of
    temperature on the mitotic cycle of normal and syn-
    chronized mammalian cells. In Cell Synchrony -
    Studies in Biosynthetic Regulation. I.L. Cameron and
    G.M. Padilla. Editors. Academic Press, Inc. New York.
    332-352.
12. Rao, P.N., and R.T. Johnson. 1970. Mammalian cell
    fusion: studies on the regulation of DNA synthesis and
    mitosis. Nature (Lond.) 225:159-164.
13. Rao, P.N., B. Wilson, and T.T. Puck. 1977. Premature
    chromosome condensation and cell cycle analysis. J.
    Cell Physiol. 91:131-142.
14. Tsui, Y.-C., M.R. Creasy, and M.A. Hulten. 1983. The
    effect of the male contraceptive agent Gossypol on
    human lymphocytes in vitro: traditional chromosome
    breakage, micronuclei, sister chromatid exchange, and
    cell kinetics. J. Med. Genet. 20:81-85.
15. Wang, Y.C., and P.N. Rao. 1984. Effect of gossypol
    on DNA synthesis and cell cycle progression of mam-
    malian cells in vitro. Cancer Res. 44:35-38.
16. Ye, W.-S., J.C. Liang, and T.C. Hsu. 1983. Toxicity
    of a male contraceptive, gossypol, in mammalian cell
    culture. In Vitro 19:53-57.

GROWTH ACTIVATION IN ADULT RAT HEPATOCYTES

N.L.R. Bucher[1], W.E. Russell[2] and J.A. McGowan[2]

[1]Department of Pathology, Boston University School of
Medicine, Boston, MA 02118.
[2]Childrens Service, Shriners Burns Institute and Mass.
General Hospital, Boston, MA 02114.

The hepatocytes in normal adult rats are in a state
of quiescence, or G_0. They retain the capacity to
proliferate, however, and re-enter the cell cycle in
partial synchrony, responding vigorously to 2/3 hepatectomy
and slightly less so to manipulations of the diet,
administration of hormones, growth factors, toxic agents,
and other substances. The precise physiological signals
that regulate this growth remain elusive, but evidence
from a number of laboratories indicates that blood-borne
substances are responsible (1).

Hepatocyte cultures offered a means of addressing
this problem in ways not possible through whole-animal
experimentation. We considered that hepatic parenchymal
cells maintained in short term primary cultures (i.e.
during the first wave of growth) would be more likely to
reflect physiologic behavior in vivo than cells maintained
outside of the body for longer periods during which they
would become altered by adaptation to their in vitro
environment. When these hepatocytes, isolated as
described, were cultured in an artificial, serum-free
medium, non-parenchymal cells were largely absent (2,3).
Only a low percentage of the hepatocytes replicated their
DNA and almost none progressed to mitosis (2). Both of
these activities were abundantly manifested, however, when
high concentrations of intermediates of carbohydrate
metabolism were added in combination with either certain

169

purified hormones and growth factors, or with normal rat
serum (4,5,6). This responsiveness of the culture system
to a variety of growth signals opened the way to further
probing.

Although translation of cell behavior in culture into
physiological terms would be an oversimplification, our
observations suggest that information with in vivo
relevance is obtainable from in vitro studies.

METHODS

Hepatocytes were isolated by collagenase perfusion
from the livers of approximately 200 gram male Sprague-
Dawley rats, and cultured in modified Waymouth's MAB 87/3
medium, as previously described (2,7).

Normal and platelet-poor rat sera (NRS and ppNRS)
were prepared from citrated blood, also as described (8).
Rat platelets were isolated from citrated blood by repeated
centrifugations and washings with buffered saline, and
lysed by several freeze thaw cycles or by sonication. The
resulting lysate (RPL) was clarified by centrifugation, and
stored at minus 20°. Human platelets from outdated frozen
blood-bank stock were treated similarly.

DNA synthesis was determined by incorporation of [3]H-
thymidine during the final 24 hours in 3-day cultures
(2,7).

RESULTS

Because of the accumulated evidence that humoral
agents regulate liver growth in vivo, we began to explore
the effects of serum in hepatocyte cultures. We found that
dialyzed human, horse, mouse, or bovine (fetal, newborn or
calf) sera whose activities were all similar, only modestly
stimulated ^{3}H-thymidine incorporation into DNA, whereas
dialyzed normal rat serum exceeded the potency of these
others by 2-3 fold. Although extensive dialysis of serum is
generally thought to remove small molecules and
polypeptides such as insulin, this treatment did not
significantly alter the activity of the rat serum (7).

The mitogenic potency of serum, at least for various

types of cells of mesenchymal origin, is now recognized to derive to a considerable extent from the blood platelets. Of the several growth factors reported to occur in platelets, only human "platelet-derived growth factor", or PDGF, has been definitively characterized and purified to homogenity. This substance, however, appears to be generally ineffective in epithelioid cells (8) consonant with our finding that DNA synthesis in hepatocyte cultures was the same in either human or mouse serum whether or not the platelets had been removed. On the contrary, as shown in Table 1, platelet removal from normal rat serum reduced its activity by 50% and recombination of rat platelet lysate with platelet-poor rat serum restored the activity.

We partially characterized the active RPL component, employing ion exchange (CM Sephadex) and gel filtration (Agarose Biogel A 0.5M) chromatography, and found it to resemble human PDGF in being a strongly cationic, trypsin-sensitive, polypeptide-like substance, whose mitogenic activity was destroyed by sulfhydryl reducing agents. It differed from PDGF, however, in having an apparently larger

Table 1. Effects upon hepatocyte DNA synthesis of normal and platelet-poor sera with and without added rat platelet lysate or purified human platelet-derived growth factor (PDGF).

	DNA dpm x 10^{-3}/µg		
Additions:	None	RPL[b]	PDGF[c]
ppHS[a]	4,689	44,447	14,351
	±89	±1,173	±1,490
ppNRS[a]	17,354	46,169	15,176
	±1,957	±7,846	±1,406
NRS[a]	37,274	--	--
	±4,875		
None	4,998	--	6,257
	±604		±291

[a] Normal and platelet-poor rat serum (NRS and ppNRS and platelet-poor human serum (ppHS) at 5 percent concentration.   [b] Rat platelet lysate (RPL), 100 µg of protein/ml of medium.   [c] Human platelet-derived growth factor (PDGF), purified 10^5-fold, 3 µg/ml (Gift of Dr. C.D. Stiles).

molecular size and considerably greater heat and acid
lability. Whereas PDGF withstands exposure to 100° (8),
the activity of RPL is destroyed at a much lower
temperature (Table 2).

Even more pronounced differences were evident upon
comparison of the biological properties of RPL and PDGF.
In hepatocyte cultures RPL was augmented equally by
platelet-poor rat or human serum, wheres highly purified
human PGDF was totally inactive under the same conditions
(Table 1).

It has been established that PDGF causes cultured
cells to become "competent", but they do not proceed
through the cell cycle unless "progression factors" are
also present. Somatomedin C serves as such a necessary
progression factor, and serum from hypophysectomized rats
which lacks this growth hormone-dependent substance, fails
to potentiate mitogenesis in PDGF stimulated cells (8). We
found that unlike PDGF, RPL stimulation of hepatocytes was
promoted equally by sera from either normal or
hypophysectomized rats at several concentrations. Figure 1
verifies this observation at 5% serum levels, and shows in
addition that the activity of the platelet-poor serum from
hypophysectomized rats (ppHypoxRS) is the same, whether
derived from portal venous (PV) or inferior vena caval
(IVC) blood. Assay of the hypophysectomized rat serum for

Table 2: Effect of heat treatment on RPL activity, alone
or in combination with ppNRS

DNA dpm x $10^{-3}/\mu g$

RPL Added	None	Unheated	(65°,30 min)	(100°,10 min)
ppNRS (5%)	18,357 $\pm 1,342$	--	--	--
RPL (100µg/ml)		20,886 ± 217	6,313 $\pm 1,849$	9,103 ± 747
ppNRS +RPL		49,037 $\pm 3,172$	15,454 $\pm 1,503$	20,929 $\pm 1,205$
Medium only	5,940 ± 601			

Figure 1:   Comparison of ppNRS and ppHypoxRS in augmenting
RPL stimulation of hepatocyte DNA synthesis.

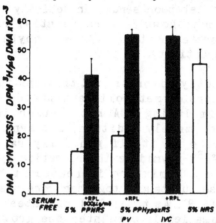

Somatomedin C content showed it to be negligible. Hence
the amplification of RPL activity by platelet-poor rat or
human serum appears not to depend upon its Somatomedin C
content.

        Despite these differences from human PDGF, it could
be argued that the RPL factor was a rat form of PDGF,
merely differing in several properties from the human form.

Table 3:   Augmentation of hepatocyte DNA synthesis by
pyruvate or lactate when combined with NRS, ppNRS or RPL.

	DNA dpm x 10^{-3}/µg		
Substrate conc.	0	2mM	20mM
10% NRS + pyruvate	19,045 $\pm$1,129	34,435 $\pm$6,140	60,362 $\pm$5,916
10% NRS + lactate	19,045 $\pm$1,129	41,930 $\pm$906	63,443 $\pm$5,705
5% ppNRS + pyruvate	15,188 $\pm$708		38,933 $\pm$2,361
RPL (100 µg/ml) + pyruvate	20,408 $\pm$3,403		39,103 $\pm$1,541

Exposure of RPL to temperatures known to destroy its
hepatocyte stimulatory capacity (Table 2), however, failed
to destroy its ability to stimulate 3T3 cells in the
presence of platelet-poor serum, an activity characteristic
of PGDF (data not shown). Consequently, rat platelets
contain two separate entities: A hepatocyte-stimulating
and a PDGF-like activity.

We previously demonstrated that the interaction of
high concentrations of metabolic substrates with insulin
and EGF in stimulating DNA replication in hepatocyte
cultures was not merely additive, but synergistic. For
example, pyruvate by itself generally raised the low,
baseline rate of ^{3}H-thymidine incorporation by several fold
and the insulin-EGF mixture increased it by 5-10 fold,
whereas all these substances together more than doubled the
additive value(6,9). As table 3 illustrates, pyruvate and
lactate also dramatically augmented the growth-promoting
substances in NRS, and pyruvate amplified the effects of
both RPL and ppNRS in similar fashion. Thus these
nutrient-growth factor interactions appear not to be
restricted to any one growth factor.

DISCUSSION

In view of the previously mentioned evidence
supporting control of liver growth by humoral signals, we
undertook studies of freshly isolated hepatocytes in
culture in order to explore effects of growth promoting
substances in serum upon the hepatocytes directly, without
intervention of other cell types. The actions of these
substances could be observed both separately, and in
combination. A serum-free culture system was estalished in
which normal adult rat hepatocytes were highly responsive
to growth stimulation, especially by insulin, EGF and
pyruvate. We employed this combination for comparative
evaluation of other putative regulators, whether
stimulatory or inhibitory.

In this system rat serum was of particular interest
because its growth-promoting potency for rat hepatocytes
considerably exceeded that of other species. The finding
that half of this activity is released from the platelets
and resides in a substance that is not PDGF, raises the
question of what, if any, physiological function it may
serve. The stimulatory components in the platelet-poor

portion of both rat and human serum are still undefined, but probably include hormones and other growth factors. As serum components may fulfill certain growth requirements peculiar to cells in culture, and in so doing obscure the actions of physiological growth regulators, resolution of the problem must await clear definition of the actions of the classes of molecules involved.

In addition to insulin, EGF, metabolic substrates and the undefined serum components described in the present study, several other purified hormones and growth factors have been reported to influence liver growth. These include catecholamines, glucagon, vasopressin, Somatomedin C, and other substances, combinations generally being more effective than substances acting singly (1, 10-12). This suggests that a similar interplay of factors may regulate hepatocyte proliferation in vivo. The existence of counteractive or inhibitory influences exerted by negative control elements remains unsettled. It seems possible that more than one combination of growth regulating agents may exercise control depending upon circumstances. The in vivo and in vitro evidence both suggest that the the metabolic state of the liver cells at the time the growth stimulus is applied is an important determinant of the particular growth promoters to which the hepatocyte will respond.

Although the precise identity of the physiological factors that regulate liver growth remains to be firmly established, the broad outlines of control by key combinations of interacting signals are beginning to emerge.

Acknowledgment: We thank Dr. C.D. Stiles for the generous gift of purified human PDGF. This work was supported by DHHS Grants No. CA 02146, AM 19435, and NICHD 3932 HD0709205, and No. 15851 from the Shriners Hospital for Crippled Children.

References:*

1. Bucher, N.L.R., McGowan, J.A. and Russell, W.E. 1983. Control of liver regeneration: present status. In Nerve, Organ and Tissue Regeneration: Research Perspectives. F.J. Seil, editor. Academic Press, New York, 455-469.
2. McGowan, J.A., Strain, A.J. and Bucher, N.L.R. 1981.

DNA synthesis in primary cultures of adult rat hepatocytes in a defined medium: Effects of epidermal growth factor, insulin, glucagon and cyclic-AMP. J. Cell. Physiol. 108:353-363.

3. Bissell, D.M. 1983. Hepatocellular function in culture: The role of cell-cell-interaction. In Isolation, Characterization and Use of Hepatocytes. R.A. Harris and N.W. Cornell, editors. Elsevier Biomedical, New York. 51-59.

4. Hasegawa, K. and Koga, M. 1981. A high concentration of pyruvate is essential for survival of DNA synthesis in primary cultures of adult rat hepatocytes in a serum-free medium. Biomed. Res. 2:217-221.

5. Hasegawa, K., Watanabe, K. and Koga, M. 1982. Induction of mitosis in primary cultures of adult rat hepatocytes under serum free conditions. Biochem. Biophys Res. Comm. 104:259-265.

6. McGowan, J.A. and Bucher, N.L.R. 1983. Hepatotropic activity of pyruvate. In Isolation, Characterization and Uses of Hepatocytes, R.A. Harris and N.W. Cornell, editors. Elsevier Biomedical, New York. 165-170.

7. Strain, A.J., McGowan, J.A. and Bucher, N.L.R. 1982. Stimulation of DNA synthesis in primary cultures of adult rat hepatocytes by rat platelet-associated substance(s). In Vitro. 18:108-116.

8. Scher, C.D., Shepard, R.C., Antoniades, H.N. and Stiles, C.D. 1979. Platelet-derived growth factor and the regulation of the mammalian fibroblast cell cycle. Biochim. Biophys. Acta. 560:217-241.

9. McGowan, J.A. and Bucher, N.L.R. 1983. Pyruvate promotion of DNA synthesis in serum-free primary cultures of adult rat hepatocytes. In Vitro. 19:159-166.

10. Leffert, H.L., Koch, K.S. Moran, T. and Rubaclava, B. 1979. Hormonal control of rat liver regeneration. Gastroenterology. 76:1470-1482.

11. Hasegawa, K. and Koga, M. 1977. Induction of liver cell proliferation in intact rats by amines and glucagon. Life Sci. 21:1723-1728.

12. Russell, W.E. and Bucher, N.L.R. 1983. Vasopressin modulates liver regeneration in the Brattleboro rat. Am. J. Physiol. 245:G321-G324.

* A full report of the work on serum and platelet factors by W.E. Russell, J.A. McGowan and N.L.R. Bucher will appear in the J. Cell. Physiol.

ISOLATION AND CHARACTERIZATION OF A GROWTH REGULATORY

FACTOR FROM 3T3 CELLS

John L. Wang and Yen-Ming Hsu

Department of Biochemistry
Michigan State University
East Lansing, MI 48824

The mouse fibroblast line 3T3 (1) exhibits a form of growth control in vitro in that it reaches only a very low saturation density and can remain for long periods of time in a viable but nondividing state. This phenomenon, termed density-dependent inhibition of growth (2), was first described and most extensively studied in the 3T3 cell line. In this paper we review some recent evidence indicating that endogenous growth inhibitors may account for at least part of the growth inhibition observed in 3T3 fibroblast cultures and that such inhibitors may be isolated and characterized in terms of their functional interactions with target cells. The results of these studies suggest several useful refinements in our analysis of the mechanisms of growth control in cultured cells.

Effect of Conditioned Medium on the Proliferative Properties of Target Cultures

Treatment of sparse, proliferating cultures of 3T3 cells with medium conditioned by exposure to density-inhibited 3T3 cultures (CM) resulted in an inhibition of growth and division in the target cells when compared with similar treatment with unconditioned medium (UCM) (3). DNA synthesis in CM-treated cells was markedly inhibited when compared with synthesis in parallel cultures treated with UCM (Fig. 1a). These results were confirmed by comparing the number of cells in target cultures treated with UCM and CM

177

(Fig. 1b). Target cells exposed to UCM continued to proliferate up to the characteristic saturation density (5 x
10^4 cells/cm^2). In contrast, cultures exposed to CM
showed much smaller increases in their cell numbers. All
of these results suggest that medium conditioned by exposure to density-inhibited 3T3 cells may contain a growth
inhibitory activity.

We have carried out other experiments to show that this
growth inhibitory activity in CM, prepared and tested in
the 3T3 system, has the following key properties: (a) It is
not cytotoxic and its effects on cell growth are reversible;
(b) The inhibitory activity can be accumulated in the medium before the onset of extensive cell-to-cell contact; (c)
The inhibitor has a more pronounced effect on target cells
at high density than on cells at lower density; (d) The
activity can be collected in the absence of serum and can
be demonstrated despite the presence of freshly added
serum; (e) The inhibitory factor(s) can be concentrated and
fractionated by precipitation and gel filtration, respectively (3).

Fractionation of the Growth Inhibitory Activity in Conditioned Medium

Gel filtration of the ammonium sulfate precipitate of CM
on a column of Sephadex G-50 partitioned the inhibitory
activity into two major components (components A and C;
Fig. 2a). Component A, which was eluted at the void volume
of the column, contained the major portion of the protein

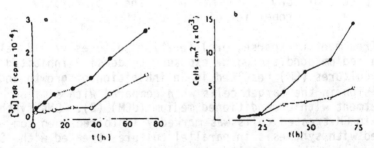

Fig. 1. Effects of CM (o) and UCM (•) on the growth of 3T3
cells. (a) Kinetics of ^{3}H-thymidine incorporation; (b)
kinetics of the increase in cell density.

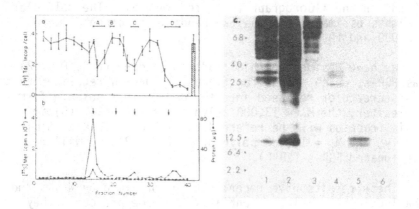

Fig. 2. Analysis of the growth inhibitory activity of CM by gel filtration on a column (90 x 1 cm) of Sephadex G-50 and by gel electrophoresis. (a) Assay of the inhibition of ^{3}H-thymidine incorporation in target cells. The vertical hatched bar at the right represents the level of DNA synthesis in target cells treated with UCM. (b) Assay of protein content by the Lowry method (●) or by counting radioactivity due to ^{35}S-methionine (o). (c) SDS-polyacrylamide gel electrophoresis of: (1) serum-free CM; (2) ammonium sulfate precipitate fraction of CM; (3) component A; (4) component B; (5) component C; (6) component D. Numbers at the left indicate the molecular weights of protein standards.

of the sample (Fig. 2b). In contrast, the second component of growth inhibitory activity, component C, was associated with a minute amount of protein material (Fig. 2b). The position of elution of component C on Sephadex G-50 suggested that the material contained polypeptide chains with molecular weights of approximately 12,000 (4).

The materials containing growth inhibitory activity at various stages of fractionation were subjected to analysis by SDS-polyacrylamide gel electrophoresis. After electrophoresis, the gel was stained with Coomassie blue and then subjected to fluorography to reveal ^{35}S-labeled protein components. The gel revealed a large number of proteins, as detected by fluorography, in CM, ammonium sulfate percipitate, and component A of the Sephadex G-50 column (Fig 2c, lanes 1-3). In contrast, component C yielded two major

bands on the fluorograph (Fig. 2c, lane 5). The molecular
weights estimated for the two bands in component C were
13,000 and 10,000.

We have designated this fraction (component C, Fig. 2)
as FGR-s, which stands for fibroblast growth regulator that
is secreted or released into the medium in a soluble form.
Hereafter, the M_r = 13,000 polypeptide (pI ~ 10) (5) in
this fraction will be referred to as FGR-s (13 K). Similarly, the M_r = 10,000 polypeptide (pI ~ 7) (5) will be
designated FGR-s (10 K).

These results have recently been confirmed by Wells and
Mallucci (6). Using essentially the same protocol, they
have partially purified a growth inhibitory activity from
medium of secondary cultures of mouse embryo fibroblasts.
The molecular weights of the polypeptides identified in the
active fractions were 13,000 and 10,000.

Identification and Neutralization of a Growth Inhibitory Factor by a Monoclonal Antibody

Using FGR-s as the immunogen, we have carried out in
vitro immunization of rat splenocytes and have generated
hybridoma lines secreting antibodies directed against components of the FGR-s preparation (7). The supernatants
from two hybridoma lines, designated 3C9 and 2A4, showed
positive reactions when assayed with either FGR-s or whole
3T3 cells (Figs. 3A and 3B). Quite the opposite results
were obtained with hybridoma clones which were derived from
rats immunized in vivo with whole 3T3 cells and screened
and selected on the basis of binding to whole 3T3 cells. A
representative clone, designated 104, is used here for
illustrative purposes. Clone 104 showed strong reaction
(2.4-fold) when assayed on whole 3T3 cells but negligible
reaction when assayed on FGR-s (Figs. 3A and 3B). These
results suggest that the products of clones 2A4 and 3C9
were monoclonal antibodies directed against some component
of FGR-s, which in turn is a constituent of whole 3T3 cells
(7).

Antibody 2A4, the rat IgG purified from clone 2A4, bound
specifically FGR-s (13 K)(Fig. 4A). There was no binding
of FGR-s (10 K). To test the possibility that Antibody 2A4
can neutralize the growth inhibitory activity of FGR-s, the

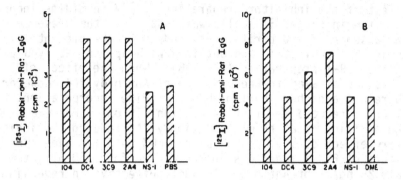

Fig. 3. Summary of the results of screening assays carried
out using (A) FGR-s and (B) whole 3T3 cells as targets. The
supernatants from clones derived from in vitro immunization
of FGR-s (DC4, 3C9, and 2A4) and from in vivo immunization
of whole 3T3 cells (104) were tested. The supernatant from
the parent myeloma line, NS-1, was used as control.

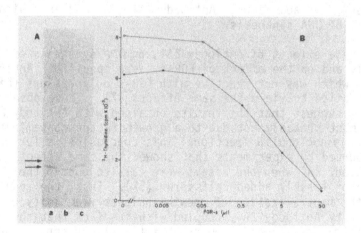

Fig. 4. Binding and functional activity of Antibody 2A4.
(A) SDS-polyacrylamide gel electrophoresis of: (a) FGR-s;
(b) FGR-s (10 K) not bound by an Antibody 2A4 affinity
column; (c) FGR-s (13 K) bound and eluted from an Antibody
2A4 affinity column. (B) Dose-response curve for the
effect of FGR-s on the incorporation of ^{3}H-thymidine by
3T3 cells in the absence (●——●) and presence (o——o) of
Antibody 2A4 (25 ng/ml).

effect of the inhibitor preparation on [^{3}H]thymidine incorporation in target 3T3 cells was assayed in the presence
and absence of the purified antibody. In the present
assay, 60% inhibition was obtained with approximately 5 µl
of the FGR-s preparation (Fig. 4B). When the effect of
FGR-s was assayed in the presence of Antibody 2A4, however,
there was a higher level of [^{3}H]thymidine incorporation
(i.e., a reduced level of growth inhibition). For example,
the addition of 25 ng/ml of Antibody 2A4 reduced the inhibitory effect of 5 µl of FGR-s from 60% to 30%. When the
amount of FGR-s added was increased ten-fold (50 µl), the
same 25 ng/ml of Antibody 2A4 was ineffective in reversing
the inhibitory effect.

Particularly striking was the observation that Antibody
2A4 (25 ng/ml) also increased the level of DNA synthesis in
3T3 cultures in the absence of any exogenously-added FGR-s
(Fig. 4B). Similarly, when a small amount of FGR-s (0.5
µl) was used, resulting in 25% inhibition, the addition of
Antibody 2A4 raised the level of DNA synthesis even beyond
that of control cultures, without any FGR-s or Antibody 2A4.
Therefore, over the entire range of FGR-s concentration
tested, the addition of Antibody 2A4 resulted in a higher
level of DNA synthesis.

These effects of Antibody 2A4, on DNA synthesis of 3T3
cells and on the effect of FGR-s, were specific. Antibody
104, which was not reactive with FGR-s polypeptides (Fig.
3), failed to yield the same effects (7). These observations suggest that the results obtained with Antibody 2A4
are most probably not due to a growth factor contaminating
the immunoglobulin fraction. This conclusion is further
supported by experiments that showed the same effects of
Antibody 2A4 when the assays were carried out in the presence of freshly added calf serum (5%). Thus, the results
on the neutralization of growth inhibitory activity of
FGR-s by Antibody 2A4, coupled with the fact that this
antibody specifically bound FGR-s (13 K), strongly suggest
that the M_r = 13,000 polypeptide carries growth inhibitory
activity.

Effect of Antibody 2A4 on 3T3 Cultures in the Absence of Exogenously-added FGR-s

If FGR-s (13 K) plays a physiologically significant role

Table I. Effect of Antibody 2A4 on DNA synthesis in 3T3 cells

Antibody 2A4 Concentration ng/ml	DNA synthesis (cpm)	Stimulation Index
0	5680	1.00
1.6	6880	1.21
8.0	9660	1.70
40.0	9150	1.61
200.0	10060	1.77
1000.0	10840	1.91

in density-dependent inhibition of growth, one would expect that addition of Antibody 2A4 to dense cultures of 3T3 cells should at least partially neutralize the activity of endogenous FGR-s (13 K) molecules in the culture and reverse the effect of density inhibition. Indeed, the addition of Antibody 2A4 resulted in a higher level of DNA synthesis in 3T3 cultures without any exogenously-added FGR-s (Table I). In contrast, the addition of Antibody 104, which is not reactive with FGR-s, failed to yield the same effect.

Antibody 2A4 binds directly to live or unfixed 3T3 cells. The binding is saturable, suggesting that there are only a finite number of antigenic targets exposed at the cell surface. To date, we have not directly identified the antigenic target of Antibody 2A4 on the cell surface. Because Antibody 2A4 specifically recognizes FGR-s (13 K), however, we infer that the $M_r = 13,000$ polypeptide or a cross-reactive precursor on the membrane may be functioning in the normal mechanism of density-dependent inhibition of growth in 3T3 cells.

Discussion

Our results can be summarized by the diagram shown in Fig. 5. Under ordinary conditions, a sparse culture of 3T3 cells in medium containing serum growth factors will be a proliferating culture. In contrast, for a dense culture (4 x 10^4 cells/cm^2), these same 3T3 cells will cease to divide as they become quiescent. FGR-s is a cellular protein fraction that will arrest sparse, proliferating 3T3 cells and convert them into a quiescent state. The $M_r = 13,000$ polypeptide in FGR-s is responsible for this action because

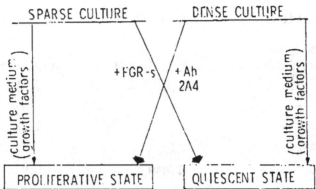

Fig. 5. Summary diagram showing the effects of FGR-s and
Antibody 2A4 on sparse and dense cultures of 3T3 cells.

the biological activity is neutralized by a monoclonal
antibody directed against the polypeptide. This monoclonal
antibody (Antibody 2A4) will also stimulate DNA synthesis
in dense, quiescent cultures of 3T3 cells. This suggests
that the target of 2A4 (FGR-s (13 K) or a cellular precursor) may be functioning in the normal mechanism of density-
dependent growth control in 3T3 cells.

The fact that FGR-s (13 K) can be isolated from soluble
conditioned medium lends support to the notion that growth
regulation at high cell densities may be partly mediated by
soluble inhibitory factors. We had previously shown that
radioactively-labeled FGR-s preparations bound specifically
to the target 3T3 cells (8). Antibody 2A4, which recognizes FGR-s (13 K), also binds to the surface of 3T3 cells
in a saturable fashion. These results suggest that FGR-s
(13 K) or a higher molecular weight precursor is present on
the plasma membrane. This in turn raises the possibility
that the endogenous growth inhibitor may also act via cell-
cell contact mediated mechanism of growth regulation, which
is supported by the early observations of 3T3 cell growth
as well as by the wound healing experiments of Dulbecco and
Stoker (2,9).

Indeed, Glaser and co-workers have shown that the growth
of 3T3 cells can be reversibly inhibited by a surface membrane fraction from the same cells and that the inhibitory
components can be solubilized by the nonionic detergent
octylglucoside (10). Alternatively, Peterson et al. (11)

have obtained similar growth inhibitory activities from 3T3
cell membranes by solubilizing the membrane with Triton
X-100 and reconstituting the solubilized components into
liposomes. Natraj and Datta have also shown that an inhibitor of DNA synthesis can be extracted from 3T3 cells by
treatment with 0.2 M urea in phosphate-buffered saline
(12). It was suggested that these surface-membrane molecules may be the same molecules that are responsible for
contact-dependent growth regulation. It would be of
obvious interest to establish the relationship between
FGR-s (13 K) and the plasma membrane associated growth
inhibitory activities. It is possible that the same molecule can exert its effects both anchored on the cell surface and released into the medium.

Acknowledgments

This work was supported by grants GM-27203 and GM-32310
from the National Institutes of Health. J.L. Wang was
supported by Faculty Research Award FRA-221 from the
American Cancer Society.

References

1. Todaro, G.J. and H. Green. 1963. Quantitative studies
   of the growth of mouse embryo cells in culture and
   their development into established lines. J. Cell
   Biol. 17:299-313.

2. Stoker, M.G.P. and H. Rubin. 1967. Density-dependent
   inhibition of cell growth in culture. Nature (Lond.)
   215:171-172.

3. Steck, P.A., P.G. Voss, and J.L. Wang. 1979. Growth
   control in cultured 3T3 fibroblasts. Assays of cell
   proliferation and demonstration of a growth inhibitory
   activity. J. Cell Biol. 83:562-575.

4. Steck, P.A., J. Blenis, P.G. Voss, and J.L. Wang.
   1982. Growth control in cultured 3T3 fibroblasts. II.
   Molecular properties of a fraction enriched in growth
   inhibitory activity. J. Cell Biol. 92:523- 530.

5. Wang, J.L., P.A. Steck, and J.W. Kurtz. 1982. Growth
control in cultured 3T3 fibroblasts. Molecular
properties of a growth regulatory factor isolated from
conditioned medium. In Growth of Cells in Hormonally
Defined Media. D.A. Sirbasku, G.H. Sato, and A.B.
Pardee, editors. Cold Spring Harbor Laboratory Press,
Cold Spring Harbor, N.Y. pp. 305-317.

6. Wells, V. and L. Mallucci. 1983. Properties of a cell
growth inhibitor produced by mouse embryo fibroblasts.
J. Cell. Physiol. 117:148-154.

7. Hsu, Y.-M., J.M. Barry, and J.L. Wang. 1984. Growth
control in cultured 3T3 fibroblasts. Neutralization
and identification of a growth inhibitory factor by a
monoclonal antibody. Proc. Natl. Acad. Sci. U.S.A., in
press.

8. Voss, P.G., P.A. Steck, J.C. Calamia, and J.L. Wang.
1982. Growth control in cultured 3T3 fibroblasts.
III. Binding interactions of a growth inhibitory
activity with target cells. Exp. Cell Res.
138:397-407.

9. Dulbecco, R. and M.G.P. Stoker. 1970. Conditions
determining initiation of DNA synthesis in 3T3 cells.
Proc. Natl. Acad. Sci. U.S.A. 66:204-210.

10. Whittenberger, B., D. Raben, M.A. Lieberman, and L.
Glaser. 1978. Inhibition of growth of 3T3 cells by
extract of surface membranes. Proc. Natl. Acad. Sci.
U.S.A. 75:5457-5461.

11. Peterson, S.W., V. Lerch, M.E. Moynaham, M.P. Carson,
and R. Vale. 1982. Partial characterization of a
growth-inhibitory protein in 3T3 cell plasma membranes.
Exp. Cell Res. 142:447-451.

12. Natraj, C.V. and P. Datta. 1978. Control of DNA
synthesis in growing Balb/c 3T3 mouse cells by a
fibroblast growth regulatory factor. Proc. Natl. Acad.
Sci. U.S.A. 75:6115-6119.

CHANGES IN ADHESION, NUCLEAR ANCHORAGE, AND CYTOSKELETON DURING GIANT CELL FORMATION

Susan Friedman, Catharine Dewar, James Thomas
and Philip Skehan
Department of Pharmacology/Oncology Research
Group, The University of Calgary, Calgary,
Alberta, Canada

Anchorage Modulation

Most cells derived from solid tissues require anchorage to a suitable surface for viability, growth, and the ability to express differentiated functions in culture (1-6). Anchorage becomes altered during growth and differentiation through changes in the contact interactions which a cell makes with other cells, substrata, and soluble extracellular ligands, and in turn influences a wide spectrum of cellular activities, including cell shape, motility, biosynthetic patterns, gene expression, and cell proliferation.

The coordination of growth, mitosis, and cell division appears to be sensitive to anchorage modulation in some but not all types of cultured cells. Cells which are highly anchorage-dependent for growth stop dividing and exhibit a partial or complete inhibition of macromolecular synthesis when cultured in suspension (7,8). If these cells are permitted to reattach to the substratum but are prevented from spreading, protein synthesis rapidly resumes but nuclear biosynthetic processes and cell division remain inhibited (9). Thus spreading, which involves the reformation of a cytoskeletal framework composed of microtubules, microfilaments, and intermediate filaments (10-13), appears to be required for reactivating DNA synthesis, RNA synthesis and processing, mitosis, and cell division (9,14).

Other types of cells fail to divide when deprived of anchorage but continue to synthesize DNA, RNA, and protein, undergo nuclear division, and transform into giant cells with variable numbers of nuclei (15). The occurrence of

187

giant cells is widespread in nature and by no means restricted to experimental situations in which anchorage is
deliberately perturbed. Giant cells are present in senesc-
ing cell populations (16), in terminally differentiating
keratinocytes (17), and throughout the culture life cycle
of a number of established cell lines. They also appear
during the development of normal and malignant tissues in
a wide variety of organisms (18-20).

The formation of giant cells by mechanisms other than
cell fusion may represent a fundamental change in cellular
growth policy that requires cells to switch from a proliferative mode, in which cell size remains fairly constant
from generation to generation, to a growth mode, in which
cell number remains constant but cell size increases. The
observation that some types of cells can be induced to make
this transition following a change in cell contact interactions suggests that anchorage may play an important role
in determining which of these modes is operative.

Giant Cell Formation in Mammalian Trophoblast

Trophoblast originates from the outermost layer of the
blastocyst (trophectoderm) and is developmentally programmed to form polyploid giant cells which serve nutritive and
hormone-producing functions during gestation. It is known
from studies on the developing mouse embryo that continued
proliferation and maintenance of diploidy of trophectoderm
requires both inductive interactions with inner cell mass
(ICM) derivatives and appropriate tissue geometry (21-23).
When trophoblast becomes separated from ICM, either experimentally or by migration during postimplantation development, it stops dividing, endoreduplicates its DNA and forms
giant cells. Studies on in vitro cultured mouse trophoblast
suggest that the arrest of cell division is not solely responsible for the induction of giant cell formation; a reduction in cell contact interactions may also be required
(23). In vivo, the formation of the blastocoel in the pre-
implantation embryo may produce shape changes in trophectoderm which serve as a stimulus for subsequent endoreduplic-
ation and giant cell formation (24).

Giant cells with greatly enlarged polyploid nuclei are
also found in human placental trophoblast and in tropho-
blastic tumors (25,26). These may arise by endopolyploidi-
zation of cytotrophoblast cells during the course of their
maturation into syncytiotrophoblast (26).

BeWo Human Choriocarcinoma: A Culture Model for Human
Trophoblast Giant Cell Formation

        The availability of a human trophoblast cell line,
BeWo (27), that can be induced to form giant syncytio-
trophoblast-like cells with high frequency in vitro (28)
provides a culture model system for studying the role of
contact interactions in the formation of giant cells and
the molecular mechanisms underlying this process. The BeWo
line was established in 1968 from a methotrexate (MTX)-
resistant human gestational choriocarcinoma (27). BeWo
cultures consist predominantly of cytotrophoblast-like
cells (CTLs) and a small proportion of giant syncytiotropho-
blast-like cells (STLs) (Figure 1). These cultures retain
many of the morphological, biochemical, and functional
properties of normal trophoblast at an early developmental
stage.

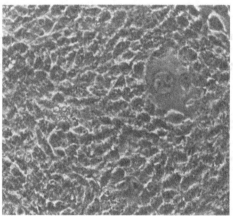

Figure 1.   Normal BeWo culture morphology

When treated with MTX concentrations that are toxic to most
cell lines (0.1-1 μM), BeWo cells stop dividing and transform into giant STLs (Figure 2).

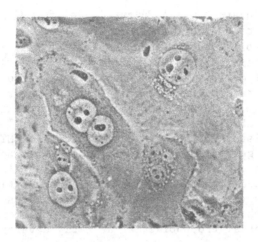

Figure 2.   STL morphology

Giant cells are first detected after 48 hours of drug
treatment. By 72 hours, more than 90% of the cells are
morphologically transformed. After 5-7 days, syncytia begin
to appear in the cultures (Figure 3).

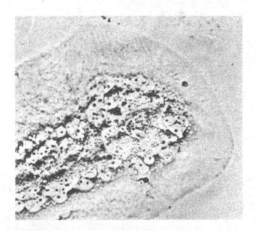

Figure 3.   Syncytium formation

        MTX produces a number of changes in BeWo cells which
affect their growth, surface and membrane properties,
interactions with neighboring cells and substrata, and
cytoskeletal organization (28-32).

Growth Characteristics:
        Soon after the addition of MTX to BeWo cultures,
cell division is arrested (Table 1). Protein and DNA
continue to accumulate for several days and cells and
nuclei become greatly enlarged.

Table 1. Inhibition of Cell Proliferation by MTX

	Millions of Cells per Flask (+/- S.D.)	
Day	Control	+ MTX
0	2.4 +/- 0.2	2.4 +/- 0.2
1	4.0 +/- 0.9	2.1 +/- 0.6
2	5.0 +/- 0.8	2.4 +/- 0.8
3	9.3 +/- 2.0	2.4 +/- 0.6

The majority of nuclei incorporate (3-H)-thymidine into
their DNA prior to and during cellular enlargement.
        During the first 24 hours of drug treatment, Feulgen-
acriflavin stained control and MTX-treated cultures have
nearly identical DNA distributions (Figure 4). Between 24
and 72 hours, there is a progressive increase in the proportion of cells with G2 and higher DNA contents, and by
day 3, the median DNA content of drug-treated cultures is
55% greater than that of the untreated cultures. Metaphase
figures are never seen in the drug-treated populations nor
has cell fusion been observed in continuous time-lapse
cinematographic recordings of cells exposed to MTX for 72
hours. These results suggest that the increase in DNA content of the drug-treated cells is produced by endoredupli-
cation.

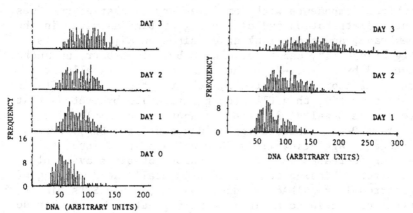

Figure 4. DNA-frequency histograms of control and MTX-
treated BeWo cultures. Frequency: number of nuclei;

sample size: 300 nuclei per culture.

Surface and Membrane Changes:
    Membrane glycosylation patterns are changed both
prior to and during giant cell formation, as determined by
(125-I)-Con A and WGA (wheat germ agglutinin) lectin
affinity staining of membrane glycoproteins fractionated on
IEF-SDS polyacrylamide gels (30). Membranes from cells
harvested at 0,24,48 and 72 hours after MTX treatment
contain deletions, additions, or changes in lectin staining in 22 Con A-reactive glycoproteins with apparent molecular weights ranging from 37-140 Kd and pIs from 4.95-8.2.
Fifty percent of the changes that appear during the 3 day
MTX treatment period are already present at 24 hours and
involve surface-localized cellular proteins. WGA-reactive
proteins are clustered into several groups with apparent
molecular weights in the 80-200 Kd range. The proteins
within each cluster are acidic (pIs<6.5), sialylated, and
differ in their isoelectric points. A distinctive change in
pattern is observed after 24 hours of MTX treatment when
one of the protein clusters disappears (MW 84Kd, pIs 4.3,
4.8,5.1,5.2,5.8) and is replaced by two less acidic species
(pIs 5.9,6.1) of the same molecular weight. It is not yet
known whether the MTX-induced changes represent new gene
products or post-translational modifications of pre-existing proteins.
    Another membrane-associated property that becomes
altered during MTX-induced giant cell formation is lectin
agglutinability (29). Cell agglutination is a complex
response that requires lectin recognition and binding to
surface components with complementary saccharide residues,
the redistribution and clustering of binding sites in the
membrane, and cross-linking of sites on adjacent cells.
Agglutinability can be regulated by cell-substratum anchorage and by density-dependent changes in extracellular
matrix factors elaborated by cultured cells (33). STLs are
less sensitive than CTLs to agglutination by lectins that
recognize D-gal ß1,3 galNAc (peanut agglutinin) and αD-
galNAc, D-gal residues (soybean agglutinin, SBA). This
change in agglutinability does not appear to involve
either decreases in the total number of sites available
for lectin binding or increases in sialic acid masking of
galactosyl or galNAc residues however, since STLs bind both
types of lectin to a significantly greater extent than do
CTLs with or without neuraminidase pretreatment (Table 2).

This suggests that the agglutinability decrease may result from changes in the mobility of lectin binding sites with the membrane.

Table 2. Binding and Agglutination of CTLs and STLs with Peanut and Soybean Lectins

		Agglutination	
Lectin	Cell Type	Aggregate[a] Ratio	(Lectin)[b] 0.5
PNA	CTL	.180	.75
	STL	.100	1.50
SBA	CTL	.185	.75
	STL	.085	3.75

Lectin	Cell Type	Lectin Bound[c] at Saturation (+) (−)	(Lectin)[d] 0.5 (+) (−)
PNA	CTL	.925 .054	25 12,22[e]
	STL	3.630 .301	25 32
SBA	CTL	3.350 .458	22 15,43[e]
	STL	5.980 .951	23 11,43[e]

[a]
Maximum agglutination expressed as the ratio of aggregates to total cells (33).
[b]
Concentration of lectin required for half-maximal agglutination, ugm/ml.
[c]
ugm lectin bound specifically per 10^6 cells with (+) and without(−)V. cholera neuraminidase pretreatment.
[d]
Concentration of lectin required for half-maximal binding, ugm/ml.
[e]
Biphasic binding curves.

Cell Contact Interactions:

Changes in cell-cell and cell-substratum interactions occur during the growth of BeWos in MTX-containing medium and may be important for the transformation of CTLs into giant STLs.

When CTLs are plated at subconfluent densities, they exhibit a marked tendency to form multicellular clusters (28). The small percentage of giant cells which are present (1-5%) tend to localize at the periphery of clusters. Moreover, the proportion of giant cells in the cultures is inversely correlated with cell density. As cultures become increasingly confluent, cells within clusters multilayer to form "dome-like" structures. Clusters and "domes" are highly resistant to dissociation by trypsin in the absence of EDTA. Fascia adherens and desmosome-like junctions are found in cell contact regions and may contribute to the stability of the clusters.

The MTX-stimulated formation of giant cells is also density-inhibited. However, unlike CTLs, STLs grow as highly flattened monolayers in which cells contact one another through extensive interdigitations of their plasma membranes and microvilli. Few desmosome-like junctions are present. The cultures are readily dispersed into single cell suspensions by trypsinization, and are attached more weakly to the substratum than are CTLs, both at subconfluent and confluent densities (Table 3). This change in cell-substratum adhesiveness occurs within 24 hours after MTX treatment and prior to the onset of morphological transformation (32).

Table 3. Differences in Adhesion of CTL and STL Cells.

Treatment	Detachment Ratio[a]
Trypsin (0.25%)/ EDTA (1mM), 25°C	.67 (subconfluent) <.22 (confluent)
DMSO (10%), 37°C	<.25
Glycine (1M)/ EDTA (2mM), 25°C	.41
EDTA (2mM), 25°C	.60

a
  Time (minutes) for 100% detachment:
  STLs/CTLs.
  Detachment assays were performed under non-shearing con-
  ditions at the indicated temperatures.

Cytoskeleton:
       The cytoskeletal system consists of a network of
protein polymers that mechanically couples the nucleus and
cytoplasmic organelles to the cell periphery and spatially
and functionally organizes the cytoplasm (34). The cortical cytoplasm plays an active role in motility-associated
functions, in cell adhesion, and in the transduction of
sensory signals (35-37). The subcortical network is
thought to be involved in cell shape formation and stabili-
zation, in the positioning and movement of organelles and
vesicles within the cytoplasm, in nuclear anchorage to the
cell periphery, and in the immobilization of polysomes,
mRNA, and cytosolic enzymes (38-43). Within the nucleus,
there exists an interchromatin matrix which has been implicated in DNA replication, gene expression, and spatial
compartmentalization (44,45).
       Cytoskeletal changes that occur during the morpho-
logical transformation of BeWo cells alter the attachment
of the nuclear-cytoskeletal framework to the substratum
(32). When cell monolayers are extracted with the nonionic
detergent Triton X-100 in the absence of microtubule sta-
bilizing agents, most of the protein and lipid is removed.
Remaining attached to the substratum is a nuclear-cytoskeletal framework that contains actin, intermediate filaments, and detergent-insoluble residues of nuclei, organelles, membranes, and pericellular matrix. The nuclear-cytoskeletal residues of CTLs and MTX-induced STLs remain
firmly attached to the substratum under these extraction
conditions, but upon further extraction with 0.3M
potassium iodide, an actin-depolymerizing agent, STL
residues are rapidly detached, whereas CTL residues remain
anchored. This change in nuclear-cytoskeletal attachment is
observed after 48 hours of MTX treatment, but not at 24
hours, when cell-substratum adhesion is affected.
       The basis for the sensitivity of STL residues to
potassium iodide detachment is not yet clear. Detached STL
and attached CTL residues appear to be morphologically
similar; each contains nuclear shells attached to and
interconnected by intermediate-type filaments. Likewise,

the two dimensional gel electrophoretic profiles of KI-
extracted residues from MTX-treated and untreated cultures
are qualitatively similar and contain the same spectrum of
major proteins, which includes several cytokeratin subunits.
However, gels of KI extractable proteins differ quantitatively and qualitatively in a number of unidentified minor
components. The most likely basis for the change in nuclear-
cytoskeletal attachment is a change in the stability,
organization, or interactions of cortical actin with intermediate filaments and membrane-associated proteins.

Rearrangements of cortical actin filaments might also
account for the changes in surface morphology which occur
during giant cell formation. Giant cells in control and
drug-treated cultures lose the surface ruffles and filo-
podia which are characteristic of CTLs and acquire numerous
short microvilli (28). Actin filaments are arranged differ-
ently in ruffles (meshworks) and microvilli (filament
bundles). These transformations may involve calcium-sensit-
ive actin binding proteins, which are demonstrably active
in cell-free systems, and are presumed to regulate actin
organization and polymerization in intact cells (46).

Changes in the organization of actin stress fibers
and cytokeratin intermediate filaments can also be visualized by indirect immunofluorescent staining of fixed
permeabilized cells with antiactin and antikeratin antibodies (Figures 5,6).

Summary and Perspectives
Considerable attention has been given in the growth
literature to the relationship between cell size and cell
division, yet somewhat surprisingly, little is known about
how cell size is regulated in division-arrested cells.

Pronounced deviations from "normal" size in individual cells within a population would be expected to produce
marked changes in their functions and interactions with
neighboring cells. In this regard, giant cells are of
particular interest insofar as cell enlargement is often
accompanied by changes in gene dosage and expression,
enhanced biosynthetic and metabolic activity, and increased
resistance to environmental damage (18,47,48). Giant cells
that remain proliferatively viable may play an important
role in cellular evolution. The division of giant cells
which have undergone rearrangements and/or changes in content of their genetic material could produce progeny with
diverse genotypes and phenotypes, and in malignant tissues,

could contribute to the rapid evolution of new species of
cells with enhanced survival capabilities in the host
(48,49).

The events that trigger giant cell formation in cells
with incomplete mitotic cycles are poorly understood. DNA
reduplication and increased transcriptional activity are
associated with the formation of giant cells in a number
of systems (18,47,49). However, it is not yet clear
whether changes in nuclear activity act as a primary stimulus for cellular enlargement or occur as a secondary consequence of other cellular changes produced by division
arrest. Likewise, giant cell formation is accompanied by
changes in membranes, cytoskeleton, adhesion and cell contact interactions (16,17,23,32), but whether these are
causally involved in the process is not known at present.
Our finding that nuclear-cytoskeletal substratum attach-
ments are altered during giant cell formation in the BeWo
system suggests that the mechanical coupling of the nucleus
to the cell periphery may itself be changed.

The mechanisms that couple nuclear function to cell
surface events are obscure. That nuclei can and do respond
to contact changes at the cell periphery is suggested by
experiments which show that nuclear biosynthetic activity
is influenced by cell spreading (9), and nuclear antigen
distribution by capping of cell surface receptors (50).
Conversely, the nucleus is required for cytoskeletal-medi-
ated functions at the cell periphery (e.g. cytokinesis
(51), ligand-induced capping of surface receptors (50,52)).
Nuclear and cell surface activities might be coupled ionic-
ally, mechanically, or by direct interactions of cell surface and cytoskeletal proteins with DNA (53,54). The
reversible binding of macromolecules to structural frame-
works in the nucleus and cytoplasm may also play an important regulatory role and must be better understood before
the complex and interesting problem of anchorage modulation
of cellular growth and function can finally be laid to rest.

Acknowledgments
This project was supported by grants from the Medical
Research Council of Canada, the Alberta Heritage Foundation
for Medical Research, and the Alberta Heritage Savings
Trust Fund-Applied Cancer Research. C.L. Dewar is a Fellow
of the Alberta Heritage Foundation for Medical Research.

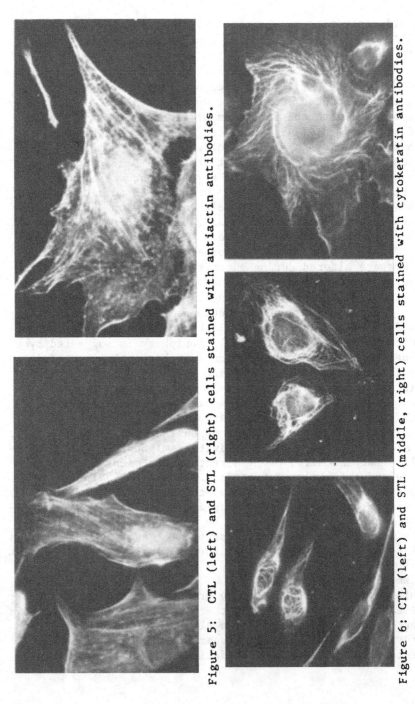

Figure 5: CTL (left) and STL (right) cells stained with antiactin antibodies.

Figure 6: CTL (left) and STL (middle, right) cells stained with cytokeratin antibodies.

References

1. Stoker, M., O'Neill, C., Berryman, S., and Waxman, V.
   Int.J.Cancer 3: 683-693 (1968).
2. Zetterberg, A. and Auer, G. Exp.Cell Res. 62: 262-
   270 (1970).
3. Maroudas, N.G. Exp.Cell Res. 74: 337-342 (1972).
4. Folkman, J. and Moscona, A. Nature 273:345-349 (1978).
5. Gospodarowitz, D., Greenberg, G. and Birdwell, C.R.
   Cancer Res. 38: 4155-4171 (1978).
6. Bissell, M.J., Hall, H.G. and Parry, G. J.Theor.Biol.
   99: 31-68 (1982), (review).
7. Otsuka, H. and Moskowitz, M. J.Cell.Physiol. 87: 213-
   220 (1976).
8. Benecke, B.J., Ben-Ze'ev, A. and Penman, S. J.Cell
   Physiol. 103: 247-254 (1980).
9. Ben-Ze'ev, A., Farmer, S.R. and Penman, S. Cell 21:
   931-939 (1978).
10. Lazarides, E. In: Cell Motility, Vol.A, (R.Goldman,
    T. Pollard, and J. Rosenbaum, Eds.), Cold Spring Harbor
    Laboratory (1976), pp. 347-359.
11. Osborn, M. and Weber, K. Exp.Cell Res. 103: 331-340
    (1976).
12. Horwitz, B., Kupfer, H., Eshhar, Z. and Geiger, B.
    Exp. Cell Res. 134: 281-290 (1981).
13. Menko, A.S., Toyama, Y., Boettiger, D. and Holtzer, H.
    Mol.Cell Biol. 3: 113-125 (1983).
14. Benecke, B.-J., Ben-Ze'ev, A. and Penman, S. Cell 14:
    931-939 (1978).
15. Ben-Ze'ev, A. and Raz, A. Cell 26: 107-115 (1981).
16. Maciera-Coelho, A. Int.Rev.Cytol. 83: 183-220 (1983),
    (review).
17. Watt, F.M. and Green, H. Nature 295: 434-436 (1982).
18. Nagl, W. Endopolyploidy and Polyteny in Differentiat-
    ion and Evolution. North-Holland Publ.Co., Amsterdam
    (1978).
19. Levan, A. and Hauschka, T.S. J.Natl.Cancer Inst. 14:
    1-20 (1953).
20. Therman, E., Sarto, G.E. and Buchler, D.A. Cancer
    Genet.Cytogenet. 9: 9-18 (1983).
21. Copp, A.J. J.Embryol.exp.Morph. 48: 109-125 (1978).
22. Gardner, R.L. J.Embryol.exp.Morph. 28: 279-312 (1972).
23. Ilgren, E.B. J.Embryol.exp.Morph. 62: 183-202 (1981).
24. Barlow, P.W. and Sherman, M.I. J.Embryol.exp.Morph.
    27: 447-465 (1972).

25. Pierce, G.B., Jr. and Midgley, A.R., Jr.   Am.J.Path.
    43: 153-173 (1963).
26. Sarto, G.E., Stubblefield, P.A. and Therman, E.   Human
    Genet. 62: 228-232 (1982).
27. Patillo, R.A. and Gey, G.O. Cancer Res. 28: 1231-1236
    (1968).
28. Friedman, S.J. and Skehan, P.   Cancer Res. 39: 1960-
    1967 (1979).
29. Skehan, P. and Friedman, S.J.   In Vitro 14: 340 (1978).
30. Friedman, S.J., Galuszka, D., Dewar, C.L. and Skehan,P.
    J.Cell Biol. 97: 52a (1983).
31. Friedman, S.J., Galuszka, D., Skehan, P., Gedeon, I.
    and Dewar, C.L. (submitted, 1984).
32. Friedman, S.J., Galuszka, D., Gedeon, I., Dewar, C.L.,
    Skehan, P. and Heckman, C.A. Exp.Cell Res (in press,
    1984).
33. Skehan, P. and Friedman, S.J. Exp.Cell Res. 92: 350-360
    (1975).
34. Cold Spring Harbor Symposia on Quantitative Biology
    XLVI, (1982).
35. Vasiliev, J.M. and Gelfand, I.M.   Int.Rev.Cytol. 50:
    159-274 (1977), (review).
36. Badley, R.A., Woods, A. and Rees, D.A.   J.Cell Sci. 47:
    349-363 (1981).
37. Edelman, G.M. Science 192: 218-226 (1976).
38. Jones, J.C.R., Goldman, A.E., Steinert, P.M., Yuspa, S.
    and Goldman, R.D. Cell Motility 2: 197-213 (1982).
39. Wang, E. and Choppin, P.W. Proc.Natl.Acad.Sci. U.S.A.
    78: 2363-2367 (1981).
40. Lehto, V.P., Virtanen, I. and Kurki, P.   Nature 272:
    175-177 (1978).
41. Cervera, M., Dreyfuss, G. and Penman, S.   Cell 23: 113-
    120 (1981).
42. Porter, K. In: Principles of Biomolecular Organization.
    (G.E.W. Wolstenholme and M. O'Conner, Eds.), Little
    Brown & Co. (1965), pp. 308-345.
43. Masters, C.J. CRC Revs.Biochem. 11: 105-143 (1981).
44. Bouteille, M., Bouvier, D. and Seve, A.P.   Int.Rev.
    Cytol. 83: 135-181 (1983), (review).
45. Hancock, R. Biol. Cell. 46: 105-122 (1982), (review).
46. Weeds, A. Nature 296: 811-815 (1982).
47. Brodsky, W. Ya and Uryvaeva, I.V. Int.Rev.Cytol. 50:
    275-332 (1977), (review).
48. Geisinger, K.R., Leighton, J. and Zealberg. J. Cancer
    Res. 38: 1223-1230 (1978).

49. Schimke, R.T.   Cancer Res. 44: 1735-1742 (1984).
50. Otteskog, P., Ege, T.E. and Sundqvist, K.G.   Exp.Cell Res. 136: 203-213 (1981).
51. Conrad, G.W. and Rappaport, R.   In: Mitosis/Cytokinesis, (A.M. Zimmerman and A. Forer, Eds.), Academic Press (1981), (review).
52. Berke, G. and Fishelson, Z. Proc. Natl.Acad.Sci. U.S.A. 73: 4580-4583 (1976).
53. Mroczkowski, B., Mosig, G. and Cohen, S. Nature 309: 270-273 (1984).
54. Traub, P., Nelson, W.J., Kuhn, S. and Vorgias, C.E. J. Biol.Chem. 258: 1456-1466 (1983).

SECTION III

TOPOLOGY OF THE CELL CYCLE

THE G1 CELL CYCLE INTERVAL IN YEASTS

R.A. Singer and G.C. Johnston*

Departments of Medicine, Biochemistry and *Microbiology

Dalhousie University, Halifax, N.S., Canada

Like most other eukaryotic cells, yeast cells regulate
the cell division process during the G1 interval of the
cell cycle (3,9). There have been two countervailing views
of the nature of G1. One view holds that the G1 interval
exists mainly to accommodate specific events after nuclear
division which are required for the subsequent S phase.
These specific events would necessarily be part of the DNA-
division sequence, a concept proposed by Mitchison (6) to
encompass periodic events in the replication and segregation of nuclear DNA. An alternative view of G1, which we
and others support (1,10), is that most of G1 is without
functional significance for the DNA-division sequence and
exists mainly to satisfy needs for ongoing processes of
mass accumulation, here termed "growth". For yeast cells
growth to what can be measured as a critical cell size is
required for performance of an event of the DNA-division
sequence prior to S phase (3,9). Clearly for these cells a
G1 will be present when sufficient mass has not accrued
during the previous cell cycle. Here we summarize experiments using the budding yeast Saccharomyces cerevisiae,
and report new results using the fission yeast
Schizosaccharomyces pombe, which show that it is solely
this dependency of the DNA-division sequence upon growth
that accounts for a G1 interval. For both yeasts,
increasing cell size at division shortens G1.

RATIONALE

Since growth and the DNA-division sequence are somewhat independent processes (but see our accompanying
report), conditions can be found which affect these two
processes in disproportionate ways. Protracting the DNA-
division sequence without proportionately affecting growth
(4,10) will delay division while the cell continues to
accumulate mass, leading to large cells. Moreover, large
cells do not have to double in mass to achieve the critical
size required for subsequent S phase initiation. If G1
exists to accommodate activities which must occur after
nuclear division but prior to S phase, these conditions
should not affect the G1 interval. If, on the other hand,
G1 exists for reasons of insufficient growth, then upon
completion of nuclear division those cells should be large
enough to be able to immediately initiate a new cell cycle.

RESULTS

Hydroxyurea (HU) causes S. cerevisiae cells to arrest
in S phase when present in high concentrations (12). At
low concentrations of HU, however, cells undergo only a
transient arrest of cell division, which then resumes at
the same rate as for an untreated culture (Fig. 1A).
During this subsequent cell proliferation the proportion of
cells without buds is decreased (4,10). Since cells in G1
are unbudded (13), this finding indicates that the length
of time spent in G1 is decreased.

For S. cerevisiae HU also affects the timing of cell
cycle events defined by certain conditional cdc mutations.
At nonpermissive temperatures cdc mutant cells accumulate
as a population of cells all arrested at one point in the
cell cycle (2). The amount of residual cell division after
temperature shift indicates the timing of a cdc step in the
cell cycle (the "execution point"). An event early in the
cell cycle, around the initiation of S phase, has an execution point of approximately 0.3. One such event is defined
by mutations in the CDC28 gene and is referred to as
"start" (3). The "start" event heralds the onset of a new
DNA-division sequence, and is the growth-dependent step in
the S. cerevisiae cell cycle (3). The execution point for
"start" is normally around 0.3. For cells dividing at normal rates in the presence of low concentrations of HU, how-

ever, the execution point for "start" is around 0.0 of the
cell cycle (4,10). This result implies that upon completion of the previous cell cycle, the CDC28 step is performed and the next cell cycle is initiated without an
intervening pre-"start" G1 interval. These findings are
not specific to HU; other ways of protracting the DNA-division sequence (Table 1) lead to similar results.

Another approach to investigate the nature of G1 has
been removal of cell cycle constraints from cells actively
dividing under conditions limiting the DNA-division
sequence. S. cerevisiae cells dividing under these conditions are unusually large (4,10). Relief of limiting conditions allows accelerated cell division for several
generations (Fig. 1B). Cells are large enough after
nuclear division to initiate a new DNA-division sequence;
no growth interval in G1 is needed and the doubling time
approximates the time required for the S+G2+M cell cycle
periods (11). During this time the execution point for
"start" remains at approximately 0.0 of the cell cycle. As
cell division outstrips growth smaller daughter cells are
produced, so that eventually normal-sized cells accumulate
with normal generation times and normal cell cycles.

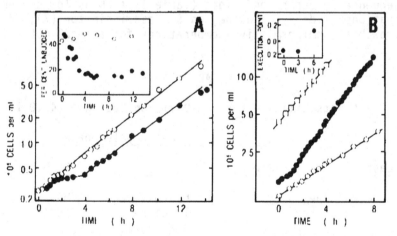

Figure 1. Effect of HU on S. cerevisiae cells. Panel A:
HU was added at time zero; from (10). Panel B: cells pregrown + HU were transferred to fresh medium ± HU at time
zero; from (11). Symbols: ● HU concentration changed at
time zero; ○, medium composition maintained; □, untreated
cells in HU-free medium.

Cells of the fission yeast S. pombe have size requirements both for initiation of S phase and for initiation of
nuclear division (9). For simplicity we have studied
mutants in which the size control over nuclear division is
functionally abolished. In these wee1 mutant cells only
the size control over S phase remains functional (9). Thus
such wee1 mutant cells of S. pombe have a cell cycle regulated similarly to that of S. cerevisiae and display a significant G1 cell cycle interval (9). As found for S.
cerevisiae, cells of S. pombe treated with low concentrations of HU were able to divide exponentially at normal
rates, but with altered cell cycle relationships. When
wee1 mutant cells were grown in limiting concentrations of
HU, the execution point for the cdc10.129 mutation marking
a pre-S phase step was shifted from 0.2 to -0.1, suggesting
that a significant time interval between nuclear division
and S phase had been eliminated. Thus in wee1 S. pombe
cells, the G1 interval appears to be present because of
insufficient growth. Cells of S. pombe growing in low concentrations of HU also were unusually large. When limiting
conditions were relieved, these large cells too displayed
accelerated rates of cell division (see our accompanying
report). Similar results were obtained (data not shown)
for wee1 S. pombe cells limited in the DNA-division
sequence by other conditions (Table 1) such as low concentrations of deoxyadenosine, an S phase inhibitor (7), or by
growth at semipermissive temperatures for a cdc22 mutant,

TABLE 1

Treatments Limiting the DNA-Division Sequence (4,10,11)

		Cell cycle phase	
Yeast	Pre-S	S	G2+M
S. cerevisiae		hydroxyurea Trenimon dTMP auxo. dUMP auxo. cdc8	MBC cdc17
S. pombe	cdc10	hydroxyurea deoxyadenosine cdc22	cdc1 cdc27

conditionally defective in replication (8). Moreover, others have found that short-term blockage of the DNA-division sequence produces large cells, and is followed by more rapid cell division (7,9).

DISCUSSION

Two approaches using two different yeasts indicate that the G1 interval, when present, is mainly an interval of growth. One approach involved protracting the DNA-division sequence. These treatments delay the completion of nuclear division, but allow mass accumulation during the slowed progress through the DNA-division sequence. Upon completion of a cell cycle, cells at or above the critical size required for initiation of the next cell cycle need no intervening period of growth. A second approach exploits the increased size of cells slowed in the DNA-division sequence. Relief of limiting conditions allows the DNA-division sequence to proceed normally, but because of the large cell sizes these cells do not have to double in mass to be large enough at nuclear division to initiate a new cell cycle. For a time these cells cycle more rapidly than normal, with a minimal G1 interval. Although there are activities normally correlated with the G1 interval, we conclude that, at least for yeast, this interval is not present to accommodate these activities, but is simply a consequence of insufficient growth during the previous cell cycle.

A similar pattern is also found in analogous HU experiments with animal cells. Growth in HU shortens the G1 interval in Chinese hamster cells (14). Conversely, slowing growth activities of strains which normally cycle without a G1 interval creates a G1 interval (5).

REFERENCES

1.   Cooper, S. 1979. A unifying model for the G1 period in prokaryotes and eukaryotes. Nature (Lond.) 280: 17-19.

2.   Hartwell, L.H., R.K. Mortimer, J. Culotti, and M. Culotti. 1973. Genetic control of the cell division cycle in the yeast. V. Genetic analysis of cdc

mutants. Genetics 74: 267-286.

3.   Johnston, G.C., J.R. Pringle, and L.H. Hartwell.
      1977. Coordination of growth and cell division in the
      yeast Saccharomyces cerevisiae. Exp. Cell Res. 105:
      79-98.

4.   Johnston, G.C., and R.A. Singer. 1983. Growth and
      the cell cycle of the yeast Saccharomyces cerevisiae.
      I. Slowing S phase or nuclear division decreases the
      G1 cell cycle period. Exp. Cell Res. 149: 1-13.

5.   Liskay, R.M., B. Kornfeld, P. Fullerton, and R.
      Evans. 1980. Protein synthesis and the presence of
      a measurable G1 in cultured Chinese hamster cells.
      J. Cell. Physiol. 104: 461-467.

6.   Mitchison, J.M. 1971. The Biology of the Cell Cycle.
      Cambridge University Press, Cambridge.

7.   Mitchison, J.M., and J. Creanor. 1971. Induction syn-
      chrony in the fission yeast Schizosaccharomyces
      pombe. Exp. Cell Res. 67: 368-374.

8.   Nasmyth, K., and P. Nurse. 1981. Cell division cycle
      mutants altered in DNA replication and mitosis in the
      fission yeast Schizosaccharomyces pombe. Mol. Gen.
      Genet. 182: 119-124.

9.   Nurse, P., and P.A. Fantes. 1981. Cell cycle con-
      trols in fission yeast: a genetic analysis. In The
      Cell Cycle. P.C.L. John, editor. Cambridge University
      Press, Cambridge. 85-98.

10.   Singer, R.A., and G.C. Johnston. 1981. Nature of the
      G1 phase of the yeast Saccharomyces cerevisiae. Proc.
      Natl. Acad. Sci. USA 78: 3030-3033.

11.   Singer, R.A., and G.C. Johnston. 1983. Growth and the
      cell cycle of the yeast Saccharomyces cerevisiae.
      II. Relief of cell-cycle constraints allows accele-
      rated cell divisions. Exp. Cell Res. 149: 15-26.

12.   Slater, M.L. 1973. Effect of reversible inhibition of
      deoxyribonucleic acid synthesis on the yeast cell
      cycle. J. Bacteriol. 113: 263-270.

13. Slater, M.L., S.O. Sharrow, and J.J. Gart. 1977. Cell
     cycle of _Saccharomyces_ _cerevisiae_ in populations
     growing at different rates. Proc. Natl. Acad. Sci.
     USA. 74: 3850-3854.

14. Stancel, G.M., D.M. Prescott, and R.M. Liskay. 1981.
     Most of the G1 period in hamster cells is eliminated
     by lengthening the S period. Proc. Natl. Acad. Sci.
     USA 78: 6295-6298.

COORDINATION OF GROWTH WITH CELL DIVISION IN YEASTS

G.C. Johnston* and R.A. Singer

Departments of *Microbiology, Medicine and Biochemistry

Dalhousie University, Halifax, N.S., Canada

During the cell cycle a cell undergoes a programmed sequence of events for replication and segregation of nuclear DNA, termed the DNA-division sequence (8). While these periodic events are taking place a cell is also increasing in size and mass. The generally continuous activities of production of new cell mass are here collectively termed "growth". Growth activities and the DNA-division sequence are normally coordinated; for yeast cells some of this coordination is manifested by growth requirements for certain cell cycle activities. In particular, for the budding yeast Saccharomyces cerevisiae, performance of the cell cycle regulatory step "start" prior to S phase requires growth to what can be measured as a critical cell size (5). For another yeast, the fission yeast Schizosaccharomyces pombe, size requirements must be met both for initiation of S phase and for initiation of nuclear division (10). During normal performance of the cell cycle in these yeasts there are periods of time during which the DNA-division sequence is temporarily held up awaiting sufficient growth to satisfy cell size requirements (5,10). Thus growth is normally the rate-limiting feature for cell proliferation.

It is generally held that growth is not regulated by the DNA-division sequence. This conclusion has been repeatedly suggested by results of experiments, using many cell systems, in which cellular growth continues when the DNA-division sequence is blocked. However, recent findings

213

using S. cerevisiae have led us to reexamine this issue.
If the DNA-division sequence is not blocked, but merely
slowed in its performance (6,12), then the DNA-division
sequence itself, not growth, becomes the rate-limiting
feature for cell proliferation. It might be expected that
under these conditions growth would outstrip the ability of
cells to divide, leading to increasingly larger progeny
cells each division. However, even though cells are
unusually large cell size does not continue to increase,
and cells display exponential growth kinetics as long as
limiting conditions are maintained. Cell viability is
unaffected (6). Such apparently balanced growth suggests
that conditions which primarily affect the DNA-division
sequence must also decrease rates of cell mass accumulation
in a parallel fashion. Experiments presented here are
designed to examine the basis for this growth limitation
when the DNA-division sequence is protracted, and suggest
that growth in S. pombe is constrained directly by the
DNA-division sequence itself.

RATIONALE

Relationships between growth and cell division have
been studied previously in the ways to be discussed here
using the budding yeast S. cerevisiae. This yeast has a
single cell size control step in its cell cycle, just prior
to S phase (5). When cells of this yeast are cultured
under conditions designed to slow but not block certain
parts of the DNA-division sequence such as S phase, those
cells show exponential but slowed rates of cell proliferation, at increased cell sizes and with altered cell cycle
parameters (6). Cell cycle analysis indicates that these
cells are likely to have a protracted S phase (6). With a
lengthened S phase, cells require a longer time to complete
the DNA-division sequence from the pre-S phase size control
step through to cell division. At nuclear division cells
are larger than normal because of the longer time for
growth from the previous size control step, and cycle with
a minimal G1 interval (6,12). When cell cycle constraints
are relieved for S. cerevisiae cells, those cells display
accelerated rates of cell division for several generations
(13). Because of initial large cell sizes those cells do
not have to double in mass to achieve the critical cell
size required for subsequent S phase initiation, and the
now-unconstrained cells, in the absence of DNA-division

sequence limitations but still unusually large, go through
several cell divisions without an appreciable G1 interval.
Smaller daughter cells are produced as cell division outstrips cell growth, so that eventually normal-sized cells
are produced (13).

For S. cerevisiae, among conditions shown to limit
cell cycle progression are the presence of hydroxyurea
(HU), and the use of semipermissive temperatures for certain cdc (cell division cycle) mutants. The S phase of the
cell cycle presumably is protracted by low concentrations
of HU because of the inhibitory effect of HU on ribonucleotide reductase (11). The rationale behind use of semipermissive temperatures, below the non-permissive temperature
where cdc gene product function is defective, is to work
where cdc gene product function is only partially impaired
(6,12). Similar limiting conditions were applied here to
cells of a different yeast, S. pombe.

Wild-type cells of the fission yeast S. pombe have two
size control steps, in different parts of the cell cycle
(10). For simplicity we have studied mutants of S. pombe
in which the operation of one size control is functionally
abolished. In these wee1 mutant cells only the cell size
control over the initiation of S phase remains functional
(10). In this way the cell cycle of wee1 S. pombe cells
resembles that of S. cerevisiae cells. Because of this S
phase size requirement, wee1 mutant S. pombe cells show a
G1 interval during which the DNA-division sequence is held
up while growth continues (our accompanying report).

RESULTS

The cell cycle of wee1 mutant S. pombe cells was
affected by growth in the presence of HU. At high concentrations HU blocks S phase in S. pombe (8), but at lower
concentrations of HU wee1 mutant cells underwent exponential cell division, at increased cell sizes and with
altered cell cycle parameters (data not shown). At an HU
concentration of 0.2 mg/ml cell division rates were
decreased but still exponential (Fig. 1A). Thus, the rate
of growth must also be decreased in parallel.

As discussed above, upon relief of conditions which
affect the DNA-division sequence, cells of S. cerevisiae

show increased rates of cell division for a time. To see
if weel cells of S. pombe behaved similarly, cells bearing
the weel.6 mutation were first grown for several generations in medium containing 0.2 mg HU/ml. As shown in Fig.
1A, when transferred to HU-free medium these cells also
displayed accelerated rates of cell division.

To examine the effects of relief of cell cycle constraints on the ability of these cells to grow we measured
accumulation of cellular protein. Cells were first grown
for several generations in the presence of both [^{14}C]-his-
tidine to label protein and HU to protract S phase. When
these cells were transferred to medium containing [^{14}C]-
histidine at the same specific activity but without HU, the
rate of growth as measured by accumulation of [^{14}C]-histi-
dine into protein was also found to increase (Fig. 1A).
Relief of constraints on the DNA-division sequence was
therefore accompanied by relief of constraints over growth

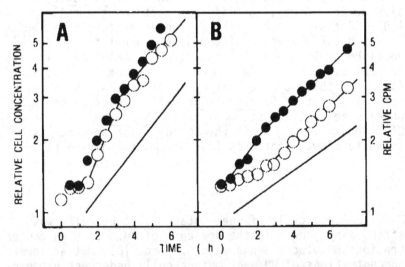

Figure 1. Relief of limiting conditions for S. pombe weel
cells. Panel A: cells pregrown in HU were transferred to
HU-free medium at time zero. Panel B: cdc 22.M45 weel.6
cells pregrown at 32.5°C were transferred to 23°C at time
zero. Symbols: O, cell concentration, and ●, [^{14}C]-histi-
dine accumulation, after relief of limiting conditions;
solid line, cell concentration in nonlimiting conditions.

rate. Moreover, the increased rate of protein accumulation
paralleled the increased rate of cell division during the
period when cells were dividing more rapidly than normal.
Evidently under these conditions cellular growth as well as
cell division can take place at rates higher than those for
cells actively dividing in the same medium but without
prior proliferation in the presence of HU.

We also used another way to alter the timing of S
phase.   A strain mutant in the cdc22 gene is defective in
DNA replication at the non-permissive temperature (9). At
the semi-permissive temperature of 32.5°C, however,
cdc22.M45 wee1.6 double mutant cells were found to proliferate exponentially, but with a generation time lengthened
by 20%. Such growing cells maintained an unusually large
cell size under these conditions. As shown in Fig. 1B,
when first grown for many generations at 32.5°C and then
returned to the permissive temperature of 23°C, cdc22.M45
wee1.6 mutant cells divided more rapidly than cells of the
same strain which had never been exposed to the higher temperature. The rate of protein accumulation also increased
upon return to 23°C, and continued at increased rates for
several hours after return of these mutant cells to the
permissive temperature. Furthermore, the growth rate shown
by these cells was greater than that of untreated cells of
the same strain in the same medium at the same temperature. These effects on growth rate and cell division rate
were not simply a result of temperature shift, since shif-
ting a cdc22[+] wee1.6 strain from 32.5°C to 23°C resulted in
an abrupt decrease in the rates of both mass accumulation
and cell division to rates characteristic of cells grown
continuously at 23°C (data not shown). It is thus likely
that the rate of performance of the DNA-division sequence,
and not the particular conditions used to limit such
performance, affects the rate at which cells can grow. The
rate of mass accumulation cannot be only a function of the
nutrient supply.

                              DISCUSSION

Results presented here suggest that for wee1 cells of
the yeast S. pombe the DNA-division sequence exerts a regulatory effect on the rate of mass accumulation. When the
rate of performance of the DNA-division sequence was
slowed, cell division was still exponential, although at

decreased rates. Under these conditions the rate of growth
must be limited in parallel with the rate of cell division. Moreover, when constraints on the DNA-division
sequence were relieved, growth increased to parallel the
increased rate of cell division. Perhaps during cell proliferation under limiting conditions "growth potential" is
"stored", and subsequently "mobilized" to allow faster-
than-normal growth after the putative growth-limiting conditions are relieved. Alternatively, growth is always
limited by the rate of performance of the DNA-division
sequence, so that faster-than-usual cell division allows
faster-than-usual growth. This latter interpretation suggests that growth rate is always less than maximal in normally proliferating S. pombe cells.

Involvement of the DNA-division sequence in growth
processes has been suggested previously for S. pombe. RNA
synthesis rates increase at some point in the cell cycle
(4), perhaps correlated with nuclear division (2,3). It
was suggested that the cell cycle controls this rate
increase (2). In addition, protein synthesis also shows
periodic rate changes during the cell cycle (1).

The experiments described here are consistent with
regulation of growth by the DNA-division sequence, but
other explanations are possible. If the conditions imposed
to limit the DNA-division sequence also themselves have
direct and parallel effects on growth, then exponential
cell proliferation could take place at decreased rates, and
relief of the limiting conditions might well allow
increased growth rates as well as increased cell division
rates. However, this explanation must account for similarity of the effects caused by both an inhibitor such as
HU and a cdc mutation with known effects on S phase.
Moreover, relief of DNA-division sequence constraints
results in a rate of mass accumulation greater than that
for untreated cells of the same strain in the same medium.

REFERENCES

1. Creanor, J., and J.M. Mitchison. 1982. Patterns of
   protein synthesis during the cell cycle of the fission
   yeast Schizosaccharomyces pombe. J. Cell Sci. 58:
   263-285.
2. Elliott, S. 1983. Coordination of growth with cell

division: regulation of synthesis of RNA during the cell cycle of the fission yeast Schizosaccharomyces pombe. Mol. Gen. Genet. 192: 204-211.

3. Elliott, S. 1983. Regulation of the maximal rate of RNA synthesis in the fission yeast Schizosaccharomyces pombe. Mol. Gen. Genet. 192: 212-217.

4. Fraser, R.S.S., and P. Nurse. 1979. Altered patterns of ribonucleic acid synthesis during the cell cycle: a mechanism for variation in gene concentration. J. Cell Sci. 35: 25-40.

5. Johnston, G.C., J.R. Pringle, and L.H. Hartwell. 1977. Coordination of growth with cell division in the yeast Saccharomyces cerevisiae. Exp. Cell Res. 105: 79-98.

6. Johnston, G.C. and R.A. Singer. 1983. Growth and the cell cycle of the yeast Saccharomyces cerevisiae. I. Slowing S phase or nuclear division decreases the G1 cell cycle period. Exp. Cell Res. 149: 1-13.

7. Mitchison, J.M. 1971. The Biology of the Cell Cycle. Cambridge University Press, Cambridge.

8. Mitchison, J.M. and J. Creanor. 1971. Induction synchrony in the fission yeast Schizosaccharomyces pombe. Exp. Cell Res. 67: 368-374.

9. Nasmyth, K., and P. Nurse. 1981. Cell division cycle mutants altered in DNA replication and mitosis in the fission yeast Schizosaccharomyces pombe. Mol. Gen. Genet. 182: 119-124.

10. Nurse, P. and P.A. Fantes. 1981. Cell cycle controls in fission yeast: a genetic analysis. In The Cell Cycle. P.C.L. John, editor. Cambridge University Press, Cambridge. 85-98.

11. Reichard, P. and A. Ehrenberg. 1983. Ribonucleotide reductase - a radical enzyme. Science (Wash., D.C.) 221: 514-519.

12. Singer, R.A. and G.C. Johnston. 1981. Nature of the G1 phase of the yeast Saccharomyces cerevisiae. Proc. Natl. Acad. Sci. USA 78: 3030-3033.

13. Singer, R.A. and G.C. Johnston. 1983. Growth and the cell cycle of the yeast Saccharomyces cerevisiae. II. Relief of cell cycle constraints allows accelerated cell divisions. Exp. Cell Res. 149: 15-26.

CHANGED DIVISION RESPONSE MUTANTS

FUNCTION AS ALLOSUPPRESSORS

Paul G. Young
Department of Biology
Queen's University
Kingston, Ontario, Canada K7L 3N6

and

Peter A. Fantes
Department of Zoology
University of Edinburgh
Edinburgh, Scotland EH9 3JT

INTRODUCTION

Schizosaccharomyces pombe adjusts cell size in response to changes in the nutritional environment.[1,2] A comparison of cell size in different media shows that cells are small under poorer growth conditions and that shifts from richer to poorer media and vice versa result in a rapid change in size. The kinetics of the response reveal that the adjustment is made at the G_2 mitotic control and is achieved by either stimulating or delaying cell division.

Three genes are known to be involved in the mitotic control: cdc 2, wee 1 and cdc 25.[3-8] The cdc 2 gene product plays a major regulatory role at mitosis. The wee 1 gene product appears to delay division, potentially by directly interacting with cdc 2 at the site of the cdc 2-w mutation. Cdc 25 functions in opposition to wee 1 and in wee 1⁻ cells it is no longer necessary for division. The molecular basis of these genetical interactions is not understood; nor is it certain where or how nutritional modulation is achieved. Wee 1⁻ cells behave as if starved (i.e.

minimal cell mass) and in fact, shifts in the execution
point for some cdc mutants are seen when starved for
nitrogen or when in a wee 1 genetic background (9;
Fantes, unpublished). Wee 1 cells do not undergo a[10]
division stimulus following nutritional shift-down
suggesting that the wee 1[+] gene product is necessary
for the functioning of the nutritional size control.

The division response to nutritional stimuli
suggested the basis of a genetic screen designed to
isolate mutants in the sensory or division
release/inhibition aspects of the mitotic control.
Changed division response mutants (cdr) were selected
for failing to undergo stimulated divisions following
an extreme shift-down (minus nitrogen); the rationale
being that this group should include cells which either[11]
did not sense or could not respond to the shift. Two
major new loci, cdr 1 and cdr 2, were identified and
all alleles have an altered response to nutritonal
shifts. They appear to have an altered perception of
their environment or their response to the nutritional
stimulus is defective.

Various data strongly suggest that nutritional
status in S. pombe is monitored as some parameter which
correlates with cell size/volume. One possibility is
that in a manner similar to that suggested in S.
cerevisiae,[12] nutritional status and hence growth rate
is perceived intracellularly at the ribosome i.e. a
monitor measuring the rate (or amount) of protein
synthesis. In S. pombe, mutants known as
allosuppressors (sal) have been found which appear to
affect the activity of nonsense suppressor tRNA's in S.
pombe.[13,14] In some cases the sal mutants have a
temperature sensitive cdc effect[13] and, in addition,
upon nitrogen starvation they tend to remain as large
cells. In addition, sal 2 was found to be an allele of
cdc 25, one of the three genes implicated in the
mitotic control (Nurse, personal communication). It
was therefore decided to test the cdr mutants for their
effect on nonsense suppression.

METHOD

The pathway for adenine biosynthesis in S. pombe
contains a step, catalyzed by the ade 6 gene product,

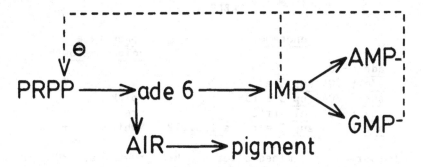

Figure 1.   Schematic of pathway for purine biosynthesis. PRPP = 5-phosphoribosyl-1-pyrophosphate; AIR = 5-aminoimidazole ribonucleotide.[15]

which, if blocked, results in the build-up of a red coloured pigment in the cells (Figure 1). Nonsense alleles of these genes have been exploited[14] to find and investigate suppressor active tRNA's. Suppression of the nonsense mutant ade 6-704 results in a shift from a red to a white coloured colony. The relative effectiveness of various suppressors can be seen as shifts in colour from red through pink to white. Sup 3-e is a serine tRNA which has an altered anticodon recognizing the nonsense codon UGA and is efficient in suppressing ade 6-704 thus making the colonies white. The tRNA can be inactivated by second site mutations within the same gene; the tRNA can then no longer function effectively as a suppressor e.g. sup 3-e,r36.[16] The allosuppressor mutations were identified as cell lines which had a third unlinked mutation causing ade 6-704 sup 3-e,r36 colonies to be white under the same growth conditions (see Table I)[13].

                    RESULTS AND DISCUSSION

    Both cdr 1 and cdr 2 were tested for allosuppressor activity by first constructing suitable strains and crossing them to generate ade 6-704 sup 3-e,r36 cdr 1-34 or cdr 2-97. A comparison of the colour and growth of various strains is shown in Table II.
    Either cdr 1 or 2 in the presence of sup 3-e,r36 affects the degree to which pigment accumulates in the

TABLE I

genotype	phenotype[1]	
	colour	growth
ade 6-704	red	−
ade 6-704 sup 3-e	white	+
ade 6-704 sup 3-e,r36	pink	−
ade 6-704 sal 3rr12	red	−
ade 6-704 sup 3-e,r36 sal 3rr12	white	+

[1]Colour: determined after 48 hr growth at 30°C on yeast extract plus 10 µg/ml adenine.
 Growth: determined in the absence of adenine on Edinburgh minimal medium.

cell. The simplest explanation is that cdr 1 and 2 are allosuppressors, facilitating the action of sup 3-e,r36 thus producing a functional ade 6 gene product and hence relieving the block to the pathway. The cells grow poorly on minimal medium showing that the block at the ade 6 step still restricts the pathway; this is a measure of the limited effectiveness of the suppression. The fact that there is some improvement in growth on minimal media eliminates most other possibilities as the cause of the reduction in pigment accumulation. If cdr mutants inhibited the adenine pathway directly, thus limiting pigment buildup, then no improved growth on minimal medium would be seen. The same is true if membrane effects allowed for an increased adenine uptake and therefore feedback inhibition of the pathway. Similarly they do not appear to allow a by-pass of the ade 6 step since there is no improvement in growth in ade 6-704 cdr 1-34 or 2-97 double mutants.

For the moment there is no basis on which to suggest a specific function for these genes. The effect on nonsense suppression could be by one of a number of routes.[14] It is unlikely that they are tRNA's themselves since they act only in concert with sup 3-e,r36; a suppressor tRNA would be expected to function alone. A second possiblity would be that they coded for enzymes involved in tRNA modification thus affecting the structure of sup 3-e,r36. In a related group of mutants, suppressor interacting, sin 1 affects

TABLE II

genotype	phenotype[1]	
	colour	growth
ade 6-704	red	-
ade 6-704 cdr 1-34	red	-
ade 6-704 cdr 2-97	red	-
ade 6-704 sup 3-e,r36	pink	-
ade 6-704 sup 3-e,r35 sal 3rr12	white	+
ade 6-704 sup 3-e,r35 cdr 1-34	white	+/-
ade 6-704 sup 3-e,r35 cdr 2-97	white	+/-

[1] see Table I

the modification of A to i^6A in tRNA.[17] Alternatively, they could represent altered ribosomal proteins thus affecting the binding properties of sup 3 on the ribosome. A general slowing of protein synthesis providing more time for the suppressor to work does not seem to be the case since cdr mutants have similar specific growth rates as compared to wild type. Lastly, it is possible that more general effects such as shifts in ionic balance could have subtle effects on codon recognition or efficiency at the ribosome.

Whatever the mechanism it is significant that the two major loci involved in preventing the cell cycle response to starvation in S. pombe both turn out to have allosuppressor activity. Similarly when sal mutants are isolated on the basis of their allosuppressor effects, one turns out to be allelic with cdc 25 (Nurse, personal communication). This is probably not coincidence. These genetic data suggest a link between the nutritional division control and the protein synthetic machinery. It seems likely that the ´growth monitor´ measures or affects some ribosomal activity and that cdr 1, cdr 2 and cdc 25 are closely associated with the mechanism and act in concert to allow division, thus acting in opposition to wee 1. This is supported by the observation that cdr 1 or 2 cdc 25 double mutants are almost lethal even at low temperatures.[11]

A number of hypothetical biochemical models could fit the essential features of the genetic and

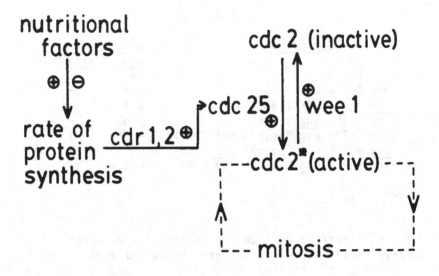

Figure 2.   Hypothetical model of mitotic control in S. pombe (see text).

physiological data.   The interrelationship of cdc 2, cdc 25 and wee 1 could be seen as an activation/inactivation cycle by assembly, modification or allosteric effects (Fig. 2).   The active state of cdc 2 is then required to allow mitosis to proceed. Wee 1 would inhibit by deactivating and cdc 25 stimulate by activating cdc 2.   The cdr 1 and 2 gene products most likely function in conjunction with the cdc 25 gene product (in this model by stimulating) since in their presence cdc 25 activity appears to be affected.   If they function in this way it could also explain the larger size of cdr 1 and 2 relative to wild type by reducing the activity of the cdc 25$^+$ gene product.   Cdr 1 and 2 would play the role of signalling the mitotic control concerning nutritional status.   In wee 1 or cdc 2-w cells cdr 1 and 2 and cdc 25 become superfluous and this is consistent with the epistatic relationship of wee 1 to them.

ACKNOWLEDGEMENTS

The authors wish to thank P. Munz for providing mutants and P. Munz and P. Nurse for providing useful suggestions and communicating unpublished data.

REFERENCES

1. Fantes, P. 1977. Control of cell size and cycle time in Schizosaccharomyces pombe. J. Cell Sci. 24: 51-67.
2. Fantes, P. and P. Nurse. 1977. Control of cell size of division in fission yeast by a growth modulated size control over nuclear division. Exptl. Cell Res. 107: 377-386.
3. Nurse, P. 1975. Genetic control of cell size at division in yeast. Nat. 256: 547-551.
4. Nurse, P., P. Thuriaux and K. Nasmyth. 1976. Genetic control of the cell division cycle in the fission yeast Schizosaccharomyces pombe. Mol. gen. Genet. 146: 167-178.
5. Fantes, P. 1979. Epistatic gene interactions in the control of division in fission yeast. Nat. 279: 428-430.
6. Thuriaux, P., P. Nurse and B. Carter. 1978. Mutants altered in the control coordinating cell division with cell growth in the fission yeast Schizosaccharomyces pombe. Mol. gen. Genet. 161: 215-220.
7. Nurse, P. and P. Thuriaux. 1980. Regulatory genes controlling mitosis in the fission yeast Schizosaccharomyces pombe. Genetics 96: 627-637.
8. Fantes, P. 1981. Isolation of cell size mutants of a fission yeast by a new selective method: characterization of mutants and implications for division control mechanisms. J. Bact. 146: 746-754.
9. Fantes, P.A. 1983. Control of timing of cell cycle events in fission yeast by the wee 1^{+} gene. Nat. 302: 153-155.
10. Fantes, P. and P. Nurse. 1978. Control of the timing of cell division in fission yeast. Exptl. Cell Res. 115: 317-329.
11. Young, P. and P. Fantes. 1984. Schizosaccharomyces mutants affected in their

division response to starvation. Submitted.

12. Unger, M. and L.H. Hartwell. 1976. Control of cell division in Saccharomyces cerevisiae by methionyl tRNA. Proc. Nat. Acad. Sci. U.S. 73: 1664–1668.

13. Nurse, P., P. Fantes, P. Munz and P. Thuriaux. 1979. Control integrating growth rate and the initiation of mitosis in fission yeast. Heredity 42: 282.

14. Egel, R., J. Kohli, P. Thuriaux and K. Wolf. 1980. Genetics of the fission yeast Schizosaccharomyces pombe. Ann. Rev. Genet. 14: 77–108.

15. De Robichon–Szulmajster, H. and Y. Surdin–Kerjan. 1971. Nucleic acid and protein synthesis in yeasts: regulation of synthesis and activity. In The Yeasts, Vol. 2. A.H. Rose and J.S. Harrison, eds. Academic Press, New York. pp. 335–418.

16. Hofer, F., H. Hollenstein, F. Janner, M. Minet, P. Thuriaux and U. Leopold. 1979. The genetic fine structure of nonsense suppressors in Schizosaccharomyces pombe. Curr. Genet. 1: 45–61.

17. Janner, F., G. Vogeli and R. Fluri. 1980. The effect of an antisuppressor on tRNA in the yeast Schizosaccharomyces pombe. J. Mol. Biol. 139: 207–219.

MOLECULAR AND CELL BIOLOGY OF CELL CYCLE PROGRESSION REVEALED BY MAMMALIAN CELLS TEMPERATURE-SENSITIVE IN DNA SYNTHESIS.

Rose Sheinin, PhD., F.R.S.C.

Department of Microbiology,

University of Toronto
Toronto, Canada M5S 1A8.

ABSTRACT

Studies of the cell cycle arrest of heat-inactivated mutant mammalian cells, which are temperature-sensitive (ts) in DNA synthesis have revealed a number of interesting processes of co-ordinated cell cycle progression. Of special relevance here are the ts A1S9, ts C1 and ts 2 mouse fibroblasts, whose mutant genes encode information for a DNA topoisomerase II enzyme, a DNA chain elongation factor and a function associated with DNA polymerase-α activity, respectively. It has been established that expression of these three defects brings cells into arrest early in S phase, in advance of the hydroxyurea restriction point, early in S phase but after the hydroxyurea execution point, and very late in G_1, at the G_1/S interface, respectively. Temperature inactivation of each unique gene product impacts in a specific way on the synthesis of the major chromosomal proteins, and on the nucleosomal structural organization of the chromatin. In addition, interruption of movement through the cell duplication cycle by heat denaturation of each mutant protein has yielded information concerning the possible involvement of cytoskeletal components in co-ordinating events of cell cycle progression.

INTRODUCTION

Some twelve years ago we turned our attention to a study of mammalian cells which are temperature sensitive (ts) in DNA synthesis (cf 1). The original goal was very simple. It was to identify those cellular enzymes/proteins which are absolutely required for the replication of the DNA of polyoma virus, and also for the cancer-producing function of this tumour virus (cf 2). In the intervening years we have moved like the Sorcerer's Apprentice (3). With every investigation of these ts cells, at least one more was generated. And even now the enchanting strains of the molecular and cell biology of somatic cells carries us onward to a fuller understanding of progression through the cell duplication cycle.

Temperature-sensitive mutants have the important advantages that they may be carried at a permissive temperature (pt) and they may be experimentally manipulated by transfer to a non-permissive temperature (npt) and back again. By such manipulation it should ultimately be possible to identify, isolate and fully characterize the polypeptide encoded by the ts genetic locus. By so doing one should expose the biochemical reaction mediated by the gene product, for subsequent complete analysis. The physiological process(es) in which the protein participates will also be revealed. And finally one aspires to isolation and full characterization of wild-type gene and its ts analogue.

Of particular importance in the present context is the potential for linking the action of a protein ts in DNA synthesis with the molecular, biochemical and physiological events of progression through the cell duplication cycle of higher eukaryotic cells.

ts Cell Families Under Study

We have now acquired four sets of rodent cells, each comprising a wild- or parental-type and one or more derivatives, which are ts in genome replication. The six mutant cells are the mouse L-cell derivatives ts A1S9 (4) and ts C1 (5); the mouse ts 2E derivative of the BalB/C-3T3 fibroblast (6); the Chinese hamster lung ts C8 isolate derived from the CHL-Ade[C] parent

(7) and the **ts** 13A and **ts** 15C CHO-EME[R] cells (8). It has been demonstrated in hybrid cell complementation tests (7,9,10) that each **ts** mutation represents a unique complementation group. Thus these **ts** cells provide us with the potential for identifying six separate enzymes/proteins of DNA replication. They further offer the possibility for exploring the impact of the six proteins on one or more processes, pathways or events of genome replication and cell cycle progression.

What is known about the ts Gene Products of Interest?

At the present time little is known about the gene products of the **ts** C8, **ts** 13A and **ts** 15C loci. In contrast we have now established that the **ts** A1S9 gene carries information for a functional DNA topoisomerase II (11, see **Table 1**). Our evidence indicates that the **ts** A1S9 gene product is a novobiocin-binding and/or sensitive peptide which is itself a component of the DNA topoisomerase II, or is an auxiliary protein essential for enzyme activity.

Table 1

THE ts A1S9 MOUSE L-CELL LOCUS

ENCODES: DNA TOPOISOMERASE II, NOVOBIOCIN-BINDING PEPTIDE

REQUIRED FOR: SEMI-CONSERVATIVE DNA REPLICATION
          MATURATION OF NEWLY-MADE SS-DNA FROM ≃5 X 10^5 TO 10^9 MOLECULAR WEIGHT

NOT REQUIRED FOR: MITOCHONDRIAL DNA REPLICATION
          DNA REPAIR REPLICATION
          POLYOMA GENOME REPLICATION
          ADENOVIRUS DNA REPLICATION
          HERPESVIRUS DNA REPLICATION
          POXVIRUS DNA REPLICATION

When we began our studies with **ts** A1S9 cells over a decade ago (12), this enzyme was deemed to be absent from eukaryotic cells. Thus our work provides the first biochemical-genetic evidence that DNA topoisomerase II is absolutely required for the replication of the genome of eukaryotic cells, by the semi-conservative mode (13).

Our experiments now make it possible to exclude
the functioning of the DNA topoisomerase II encoded
in the **ts** A1S9 locus in a number of biochemical processes.
These include the replication of mitochondrial DNA
(14), DNA repair replication (13), the replication
of the polyoma (Py) virus genome (15,16), and the
replication of the DNA of mouse adenovirus mouse cytomegalovirus and vaccina.

Table 2

THE ts C1 LOCUS

ENCODES: A DNA CHAIN ELONGATION FACTOR

REQUIRED FOR: SEMI-CONSERVATIVE CELLULAR DNA SYNTHESIS
            POLYOMA DNA REPLICATION

NOT REQUIRED FOR: MITOCHONDRIAL DNA REPLICATION
                DNA REPAIR REPLICATION
                ADENOVIRUS DNA REPLICATION
                HERPESVIRUS DNA REPLICATION
                POXVIRUS DNA REPLICATION

Although we have not yet identified the **ts** C1
gene product, we have shown that it is essential for
the nuclear replication of cellular (14,17,18) and
Py DNA (**cf** 19,20, see **Table 2**). The evidence indicates
that the **ts** C1 protein participates in the process
of polydeoxyribonucleotide chain elongation. Since
ts C1 cells are not **ts** in their DNA polymerase (α
-β) activity (cf 17,18, Philippe, Sheinin and de
Recondo, unpublished), our studies with **ts** C1 cells
may herald the biochemical-genetic identification
of an auxiliary protein or other component of a DNA
replication complex (**cf** 21). They clearly reveal
that the **ts** C1 protein is an obligatory participant
in normal semi-conservative DNA replication, but is
not required for repair replication (13). The **ts**
gene product plays no direct part in mitochondrial
DNA synthesis (14,17,18), nor in the replication of
the genomes of mouse adenovirus, mouse cytomegalovirus
or vaccinia virus.

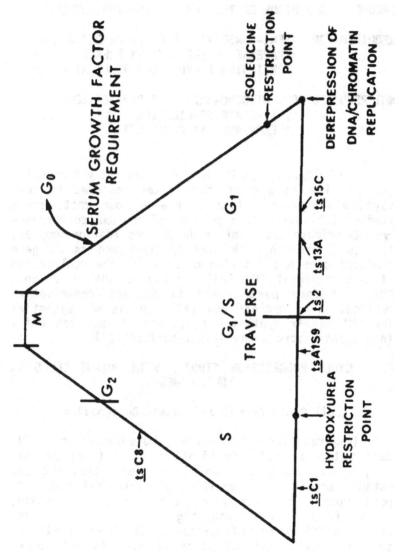

CELL CYCLE ARREST OF MAMMALIAN
CELLS *ts* IN DNA SYNTHESIS

Table 3

THE ts MOUSE CELL LOCUS

ENCODES: A PEPTIDE OF THE DNA REPLICATION COMPLEX

REQUIRED FOR: SEMI-CONSERVATIVE CELLULAR DNA SYNTHESIS
TRAVERSE OF THE G_1/S INTERFACE
CONTINUING POLYOMA DNA REPLICATION

NOT REQUIRED FOR: MITOCHONDRIAL DNA REPLICATION
DNA REPAIR REPLICATION
ADENOVIRUS DNA REPLICATION

The BalB/C-3T3 ts 2E cell also has not yet revealed the precise nature of the protein encoded in this particular genetic locus. However our most recent studies (22) point to a peptide of the DNA polymerase-α: DNA primase component of the DNA replication complex. The ts 2 protein, like the ts A1S9 and ts C1 gene products does not function directly in the replication of mitochondrial DNA (23), adenovirus DNA or polyoma DNA. It does participate in the semi-conservative replication of nuclear DNA (23), but is not essential for DNA repair synthesis (23). And it mediates entry into S phase across the G_1/S interface (23).

CELL CYCLE PROGRESSION STUDIED WITH MUTANT CELLS ts IN DNA SYNTHESIS

Cell Cycle Arrest and Mutant Designation

Temperature-sensitive mutants have been pragmatically designated as cell duplication cycle (cdc) mutants (24,25) and non-cdc mutants. By definition the cdc mutants arrest at a specific, physiological execution point upon temperature inactivation of the ts protein. As a first step in studying the cell cycle impact of ts mutations in DNA synthesis, we first identified the cell cycle arrest point of each ts cell under study. These are indicated in Figure 1. Because the ts A1S9 and ts C1 cells arrest in S phase, these are designated as dna^{ts}/S^{ts}. In contrast the ts 2E cells reveal an execution point very late in G_1, or very near the G_1/S interface. They are denoted as dna^{ts}/G$_1$^{ts}.

The temporal map of cell cycle progression set out in **Figure 1** can only be considered in terms of the experimental operations used to define the cell cycle (26). However some greater confidence in such framing of the cell duplication cycle derives from the recognition that the **cdc ts** mutants obtained are in fact compatible with the procedures used for their generation and selection (**cf** 9,10,27).

The several protocoles employed should have generated a large number of cells mutant in a function of G$_1$. However they would also have yielded cells **ts** in DNA replication due to mutant enzymes/proteins which act very late in G$_1$, at the G$_1$/S interface or very early in S phase. These predictions are made somewhat imprecise by our own demonstration (**cf** 1,9,17,18,28) and that of others (27), that **ts** cells generally, and certainly those which are **ts** in DNA synthesis, lose their capacity for cell division and concomitant cell cycle progression within one or two generation period equivalents after upshift to the **npt**. This undoubtedly accounts for the observation that very few **dna^{ts}/S^{ts}** and **ts** 2-like mutant cells have been isolated to date (**cf** 1). In contrast G$_1$^{ts} mutants are quite numerous (29,30).

Cell Cycle Arrest: Relevance of ts Gene Product

It was noted earlier that **ts** mammalian cells have been loosely classified as **cdc** or non-**cdc** mutants, on the basis of whether or not they arrest at a specific stage of the cell cycle upon temperature inactivation. It is of interest for the validity of such a parameter for **ts** mutant classification, to examine the nature and function of **ts** gene products once identified.

Table 4

THE ts A1S9 LOCUS: CELL CYCLE BLOCK

GENE PRODUCT: DNA TOPOISOMERASE II

CELL CYCLE ARREST: EARLY S PHASE

BLOCK OF de novo DNA SYNTHESIS AT ˜5 X 10^8 (SS)
CO-ORDINATE SHUT-DOWN OF HISTONE SYNTHESIS
CO-ORDINATE SHUT-DOWN OF SYNTHESIS OF OTHER CHRO-
MOSOMAL PROTEINS
DISAGGREGATION OF FACULTATIVE, CONDENSED HETERO-
CHROMATIN OF THE NUCLEOPLASM
MODIFICATION OF THE SUPERCOILING OF THE CHROMATIN
DNA (EtBr-BINDING: INCREASED DNase I SENSITIVITY)

It is both interesting and gratifying to discover
that the ts A1S9 (cf 1) and ts C1 (17,18) cells, which
do arrest in S phase are ts in proteins which have
been demonstrated to participate directly in the process
of nuclear DNA replication.

Table 5

THE ts C1 LOCUS: CELL CYCLE BLOCK

GENE PRODUCT: DNA CHAIN ELONGATION FACTOR

CELL CYCLE ARREST: EARLY S PHASE, BEYOND ts A1S9

CO-ORDINATE SHUT-DOWN OF HISTONE SYNTHESIS
CO-ORDINATE SHUT-DOWN OF SYNTHESIS OF OTHER CHRO-
MOSOMAL PROTEINS
DISAGGREGATION OF FACULTATIVE CONDENSED HETERO-
CHROMATIN OF THE NUCLEOPLASM
MODIFICATION OF SUPERCOILING OF THE CHROMATIN
DNA (ETBR-BINDING; INCREASED DNASE I SENSITIVITY)

The ts 2 peptide appears to be involved in entry
into the DNA-synthetic process. This is compatible
with the observation (31) that ts 2 cells arrest very,
very late in G_1 under restrictive conditions. In
fact, ts 2 cells temperature-inactivated through the
equivalent of over one generation period behave opera-
tionally as though they were blocked just in advance

of the G_1/S traverse. Thus such cells resume DNA synthesis at the normal rate almost immediately upon backshift to permissive conditions (6,31).

Table 6

THE **ts** 2 LOCUS: CELL CYCLE BLOCK

GENE PRODUCT: PEPTIDE COMPONENT OF DNA REPLICATION COMPLEX

CELL CYCLE ARREST: VERY LATE G_1, THE G_1/S INTERFACE

UNCOUPLING OF HISTONE SYNTHESIS FROM TEMPERATURE-INACTIVATED DNA SYNTHESIS
UNCOUPLING OF THE SYNTHESIS OF THE OTHER CHROMOSOMAL PROTEINS
MODIFICATION OF SUPERCOILING OF THE CHROMATIN DNA (ETBR-BINDING; DECREASED DNASE I SENSITIVITY)

In this context it is worthwhile to point out that the G_1 ^{ts} **ts** AF8 hamster cell (32) which arrests early in G_1, is mutant in a subunit of RNA polymerase II (33,34). This enzyme is essential for the transcription of those m-RNA molecules which translate into proteins known to be required for entry into active cell cycle progression at an early G_1 stage (34,35).

Impact of **ts** Gene Expression on Cell Cycle Progression

The study of **ts** cells incubated at the **npt** for varying intervals has revealed a number of interesting regulatory events associated with cell cycle progression (see **Tables 4-6**). The first involves the relationship between nuclear and mitochondrial DNA synthesis. Thus when growing **ts** A1S9 (14), **ts** C1 (17,18) or **ts** 2 (23) mouse cells are upshifted to the **npt**, replication of their mitochondrial DNA proceeds normally at the control rate throughout the first generation time equivalent, even though nuclear DNA synthesis is severely inhibited. As the cells move into what would have been a second cell cycle, the rate of mitochondrial DNA replication declines abruptly such that the ratio of synthesis in the two cellular DNA compartments is maintained.

A second regulatory process is revealed when
the **ts** A1S9, **ts** C1 or **ts** 2 cells are maintained at
the **npt** into and through a third, aborted cell cycle
under restrictive conditions. Thus we have demonstrated
that these **ts** cells continue to replicate their DNA
by the normal semi-conservative mode through the equivalent
of two generation periods, even though the rate of
nuclear DNA synthesis is greatly reduced (13,23).
As the cells move into what would have been a third
duplication cycle all of the DNA made results from
repair replication.

These events may be related to the general finding
that **ts** A1S9 (cf 1,36), **ts** C1 (cf 1,17,18), **ts** 2 (31)
and other **ts** cells (cf 1) go into unbalanced growth
at the **npt**. The synthesis of RNA and protein is apparently
normal, as indicated by the rate of incorporation
of labelled precursor. However it is clear from studies
of specific proteins that the patterns of synthesis
of these macromolecules is abnormal. In **ts** A1S9 cells,
the turnover of selected cytoplasmic proteins is also
out of the ordinary (36). Nevertheless the continued
synthesis of other protoplasmic constituents, in the
absence of DNA replication and cell division, gives
rise to greatly enlarged cells upon long-term incubation
at the **npt**.

The phenomena of unbalanced growth and switch-on
of DNA repair replication are indicative of abnormal
cell cycling of the **ts** cells, once they have been
incubated at restrictive conditions late into the
aborted second cell duplication cycle and beyond.
This hypothesis is compatible with the observed plating
efficiencies of **ts** cells incubated at the **npt**. All
show little or no capacity for cell division once
they move into the period of DNA repair replication
and major unbalanced growth. In contrast all of these
cells, maintained at the **npt** throughout the first
cell cycle, reveal essentially full plating efficiency
when they are subsequently incubated at the **npt** (cf
1,17,31).

The latter finding suggests that the **ts** cells
move normally through their first duplication cycle
at the **npt**; or attempt to do so. This postulate is
in accord with the observations that the **ts** cells

of interest carry out apparently normal, semi-conservative DNA replication throughout the first and second cell cycle intervals at the **npt**. As already noted **ts** 2 cells incubated at the **npt** for an analagous period appear to recover immediately upon backshift to the **pt**. The **ts** A1S9 and **ts** C1 cells also recover from the aborted second cycle. However the interval between backshift and onset of recovery increases throughout this cycle, as a function of the actual period of high temperature exposure (cf 1,17,18). Once the **ts** 2 cells have moved into the second aborted cell cycle period at the **npt**, recovery upon backshift to the **pt** requires a full generation period (32).

Impact of ts Gene Expression on Structural Components of the Cell

ts A1S9, **ts** C1 and **ts** 2 mouse fibroblasts subjected to heat inactivation for 20-24 hrs., i.e. for one cell cycle interval; exhibit highly synchronous growth behaviour (cf 1,9,17,18,31). This observation indicates that temperature inactivation of the specific **ts** gene product permits the cells to cycle normally to the execution point for cell cycle progression. It therefore suggests that temperature manipulation of **ts** cells should yield information about two kinds of cell cycle processes; those which continue normally while the specific **ts** reaction is undergoing inactivation and those which block and result in the accumulation of structures and proteins which act during the cell cycle in advance of the protein encoded in the particular **ts** gene.

We have observed that the **dna**^{ts}/s^{ts} **ts** A1S9 and **ts** C1 mouse L-cells undergo temperature inactivation of DNA replication and subsequent coupled termination of the synthesis of all the histones (37,38) and other chromosomal proteins (see **Table 4 and 5**). It is known that the synthesis of the DNA and histones is normally co-ordinately turned on and then turned off during S-phase (cf 39), as the chromatin undergoes duplication. We have shown that direct metabolite inhibition of DNA replication in wild-type cells gives the same coupled inhibition of DNA and histone synthesis seen in the heat-inactivated **ts** A1S9 and **ts** C1 cells (40).

On the basis of the foregoing findings we have concluded that the dna^{ts}/S^{ts} cells move normally into the DNA-synthetic phase. Temperature inactivation of each individual ts gene product interrupts ongoing chromatin DNA replication, without interfering with movement through the S phase already put in motion. Thus synthesis of chromosomal proteins continues until the machinery responds fully to the temporal signal for complete chromatin replication, initially given as the cells passed the G_1/S traverse. Thereafter the formation of the chromosomal proteins is terminated.

The dna^{ts}/S^{ts} cells appear to have revealed another process of S phase, associated with duplication of the genome housed in the chromatin proteins. This is manifest (see Tables 4 and 5), as the progressive disaggregation of the facultative, condensed hetero-chromatin of the nucleoplasm in ts A1S9 and ts C1 cells incubated through the first and into the second aborted cell cycle (41). These observations have been interpreted as the outcome of two processes. The first, initiated early in S phase, is postulated to be mediated by site-specific proteases which act to release compacted chromatin into a configuration which would receive the DNA replication complex (cf 1). The continuous, uninterrupted action of this system would ultimately yield fully decondensed chromatin. Under normal circumstances this disaggregating process would be phenomenologically reversed by the deposition of newly-made, newly-deposited histones, etc. onto the newly-replicating DNA, producing locally-compacted DNA and recondensed chromatin. In the temperature-inactivated dna^{ts}/S^{ts} cells there is no newly-made DNA. There is no substrate to receive newly-made chromosomal proteins. Hence the apparently abnormal pleiotropic phenomenon of disaggregation of chromatin seen in temperature-inactivated dna^{ts}/S^{ts} cells, bespeaks at least two normal processes of S phase progression.

The dna^{ts}/S^{ts} cells reveal yet a third general process of S phase progression. Thus temperature inactivation of the ts A1S9 and ts C1 cells is associated with modification of the supercoiling of their chromatin DNA. This is revealed (see Tables 4 and 5) in altered patterns of binding of the intercalating dye, ethidium bromide (42). It is also revealed by markedly enhanced sensitivity to the endonuclease DNase I (Restivo, Cremisi and Sheinin, unpublished data).

Recently we have turned to a study of the organization of cytoskeletal structures in temperature-inactivated dna**ts**/s**ts** cells. In collaboration with Dr. Jane Aubin (MRC Periodontal Disease Unit, University of Toronto), we have examined wild-type (WT-4), **ts** AlS9 and **ts** Cl mouse cells incubated at the **pt** or the **npt**, using probes for actin, tubulin and vimentin (**cf** 43). The results indicated no major differences in the organization and distribution of the tubulin and actincontaining elements of these cells. In contrast the distribution of the vimentin-containing intermediate filaments was markedly altered in **ts** AlS9 cells brought to full cell cycle arrest.

Almost all of the heat-blocked **ts** AlS9 cells revealed the accumulation of an usual structural network, by immunofluorescent probing with a monoclonal antibody to purified vimentin. This extra-nuclear network carried filamentous extensions which appeared to originate in an agglomeration or nucleation centre at one pole of the nucleus.

The accumulation of such vimentin-containing networks was very marked in **ts** AlS9 cells arrested at the **npt**, being present in a majority of cells. Although this cytoskeletal structure was detected in temperature-inactivated **ts** Cl cells, its accumulation appeared to be less extensive in amount per cell, and in the number of cells affected. The same kind of cytoskeletal structure was detected with very low frequency in control cultures. These observations suggest interpretation along the following lines. The intermediate filament network may serve normally in a highly dynamic process for the ordered and regulated delivery of nuclear proteins from their site of cytoplasmic synthesis to their site of nuclear action/function. Interruption of S phase progression, eg. by temperature inactivation of the **ts** AlS9 or **ts** Cl gene products, freezes the process of network formation: dissolution at a unique stage.

This model gains additional interest from very preliminary studies performed in collaboration with the laboratory of Dr. David Brown (Department of Biology, University of Ottawa), which is in the process of applying monoclonal antibody technology to the study of the function and structural organization of proteins of the nuclear matrix. This group has prepared monoclonal antibodies to nuclear matrix isolated from resting peripheral lymphocytes of mice. One of these, designated as 5H12, is of special interest in the present context. Thus this antibody reacts with an antigen which accumulates in almost every cell in a culture of temperature-arrested ts A1S9 cells, in a structure which resembles that revealed by antibody to vimentin, noted above. Although this antigen is present in all other cells examined, it does not accumulate in such networks.

These observations can be accomodated within the model enunciated above. It invokes the transient mobilization very early in S phase, of a cytoskeletal network composed of intermediate filaments which guide or interact with proteins made in the cytoplasm, but destined for a structural or catalytic function in the nucleus.

The foregoing studies with the dna^{ts}/S^{ts} cells yielded results markedly different from those with the dna^{ts}/G$_1$^{ts} cells (see Table 6). Thus in the latter, the process of synthesis of the chromosomal histones and other proteins (38) is uncoupled from temperature-inhibited DNA synthesis. The facultative condensed heterochromatin of the nucleoplasm retains its structural organization as detected by electron microscopy (31). The ts 2 cells do reveal a modification in the state of supercoiling of the chromatin-DNA as evidenced by ethidium bromide binding (31). However, whereas the dna^{ts}/S^{ts} cells arrested in S phase exhibited enhanced susceptibility to attack by DNase I, the ts 2 cell chromatin DNA presented remarkable resistance to this enzyme.

CONCLUSION AND PERSPECTIVES

The studies set out here indicate the kinds of information which can be garnered by the exploitation of mammalian cells which are ts in DNA synthesis. The key information is, of course, the nature and function of the specific gene product. Perhaps equally important is the establishment of the cell cycle execution or arrest point, which follows upon denaturation of the **ts** protein. An analysis of the several pleiotropic expressions of the primary mutation can then lead to a synthesis of the physiological, biochemical and molecular events of cell cycle progression. We have already begun to make major inroads towards elucidating the part played by the eukaryotic DNA topoisomerase II protein encoded in the **ts** A1S9 genetic locus, the DNA chain elongation factor determined by the **ts** C1 gene and the **ts** 2 protein. We look forward with anticipation to the secrets to be revealed by these and the other **dna**^{ts} mutations already available and yet to be isolated.

ACKNOWLEDGEMENTS

The moral and financial support of the following agencies is gratefully acknowledged! l'Association Pour la Recherche Contre le Cancer; l'INSERM; la Ligue Nationale Contre le Cancer; the Medical Research Council of Canada; the National Cancer Institute of Canada.

REFERENCES

1.   Sheinin, R. 1980. Mutants in the study of cell cycle progression in mammalian cells. In **NUCLEAR AND CYTOPLASMIC INTERACTIONS IN THE CELL CYCLE.** (Ed. G.L. Whitson) Academic Press, New York, pp. 105-166.

2.   Tooze, J. 1981. **THE MOLECULAR BIOLOGY OF THE DNA TUMOUR VIRUSES** Cold Spring Harbour Laboratory. Cold Spring Harbour, New York.

3.   Disney, W. 1940. **FANTASIA** Walt Disney Studios. Los Angeles California. Film interpretation of **The Sorcerer's Apprentice by** Paul Dukas.

4.   Thompson, L.H., Mankovitz, R., Baker, R.M., Till, J.E., Siminovitch, L. and G. F. Whitmore. 1970. Isolation of temperature-sensitive mutants of mouse L-cells. Proc. Natl. Acad. Sci. USA **66**: 377-384.

5. Thompson, L.H., Mankovitz, R., Baker, R.M., Till,
    J.E., Siminovitch, L. and G.F. Whitmore. 1970.
    Selective and non-selective isolation of temperature-
    sensitive mutants of mouse L-cells and their
    characterization. J. Cell. Physiol. 78: 431-440.

6. Slater, M.L. and H.L. Ozer. 1976. Temperature-
    sensitive mutants of BalB/C 3T3 cells. II. Description
    of a mutant affected in cellular and polyoma
    DNA synthesis. Cell 7: 289-295.

7. McCracken, A. 1982. A temperature-sensitive
    DNA synthesis mutant isolated from the Chinese
    hamster ovary cell line. Somat. Cell Genet. 8:
    179-195.

8. Srinivasan, P.R., Gupta, R.S. and L. Siminovitch.
    1980. Studies on temperature-sensitive mutants
    of Chinese hamster ovary cells affected in DNA
    synthesis. Somat. Cell Genet. 6: 567-582.

9. Siminovitch, L., Thompson, L.H. Mankovitz, R.,
    Baker, R.M. Wright, J.A., Till, J.E. and G.F.
    Whitmore. 1973. The isolation and characterization
    of mutants of somatic cells. Canad. Cancer Res.
    Conf. 9: 59-75.

10. Jha, K.K., Siniscalco, M. and H.L. Ozer, 1980.
    Temperature-sensitive mutants of BalB/3T3 cells. III.
    Hybrids between ts 2 and other mutant cells affected
    in DNA synthesis and correction of ts 2 defect
    by human X chromosome. Som. Cell Genet. 6: 603-614.

11. Colwill, R.W. and R. Sheinin. 1983. Evidence
    that the ts A1S9 locus in mouse L-cells may encode
    a novobiocin binding protein which is required
    for topoisomerase II activity. Proc. Natl. Acad.
    Sci. USA. 80: 4644-4648.

12. Sheinin, R. 1976. Preliminary characterization
    of the temperatuure-sensitive defect in DNA replica-
    tion in a mutant mouse L-cell. Cell 7: 49-57.

13. Sheinin, R. and S.A. Guttman. 1977. Semi-conservative and non-conservative replication of DNA in temperature sensitive mouse L-cells. Biochim. Biophys. Acta. **479**: 105-118.

14. Sheinin, R., Darragh, P. and M. Dubsky. 1977. Mitochondrial DNA synthesis in mouse L-cells temperature-sensitive in nuclear DNA replication. Canad. J. Biochem. **55**: 543-547.

15. Sheinin, R. 1976. Polyoma and cell DNA synthesis in mouse L-cells temperature-sensitive in the replication of cell DNA. J. Virol. **17**: 692-704.

16. Ganz, P.R. and R. Sheinin, 1983. Synthesis of multimeric polyoma virus DNA in mouse L-cells: Role of the **ts** A1S9 gene product. J. Virol. **46**: 768-777.

17. Guttman, S. and R. Sheinin. 1979. Properties of **ts** C1 mouse L-cells which exhibit temperature-sensitive DNA synthesis. Exptl. Cell Res. **123**: 191-205.

18. Guttman, S. 1977. **The Characterization of a temperature-sensitive mouse L-cell - ts C1.** MSc. Thesis, University of Toronto. Toronto, Ontario.

19. Sheinin, R., Colwill, R.W. and P.R. Ganz. 1983. Polyoma and cell chromatin replication studied in mouse cells which exhibit temperature-sensitive DNA synthesis because they are S^{ts} or G_1^{ts}. In **NEW APPROACHES IN EUKARYOTIC DNA REPLICATION.** (Ed. A.M. de Recondo) Plenum Press Corporation, London pp. 277-291).

20. Sheinin, R. Strategies utilized by papovaviruses to modify the genome of host cells. In **ENDOCYTOBIOLOGY II, PROCEEDINGS OF THE SECOND INTERNATIONAL COLLOQUIUM ON ENDOCYTOBIOLOGY.** (Ed. W. Schwemmler & W.E.A. Schenk). Walter de Gruyter, Berlin, 1983. pp. 69-82.

21   Cobianchi, F., Riva, S., Mastromei, G., Spadari,
      S., Pedrali-Noy, G. and A. Falaschi. 1979. Enchance-
      ment of the rate of DNA polymerase-   activity
      on duplex DNA by a DNA-binding protein and a
      DNA-dependent ATPase of mammalian cells. Cold
      Spring Harbour Symp. Quant. Biol. **48**: 639-648.

22.   Philippe, M., Sheinin, R. and de Recondo A.M. 1984.
      Biochemical genetic analysis of the mechanism
      of the G_1/S traverse in mammalian cells. In
      press.

23.   Sheinin, R., Dardick, I., Sparkuhl, J. and F.W.
      Doane. 1984. Phenotypic expressions of the **ts**
      2 mutation of BalB/C-3T3 mouse fibroblasts which
      are temperature-sensitive for DNA synthesis.
      Manuscript in preparation.

24.   Hartwell, L.H. 1976. Cell division from a genetic
      perspective. J. Cell Biol. **77**: 627-637.

25.   Pringle, J.R. 1978. The use of conditional lethal
      cell cycle mutants for temporal and functional
      sequence mapping of cell cycle events. Physiol. **95**:
      393-406.

26.   Howard, A. and S.R. Pelc. 1953. Synthesis of
      desoxyribonucleic acid in normal and irradiated
      cells and its relation to chromosome breakage.
      Heredity 6: Supp. 261-273.

27.   Basilico, C. 1977. Temperature-sensitive mutations
      in animal cells. Adv. Cancer Res. **24**: 223-266.

28.   Savard, P., Poirier, G. and R. Sheinin. 1981.
      Poly (ADP- ribose) polymerase activity in mouse
      cells which exhibit temperature-sensitive DNA
      synthesis. Biochim. Biophys. Acta. **653**: 271-275.

29.   Basilico. C. 1978. Selective production of cell
      cycle specific **ts** mutants. J. Cell Physiol. **95**:
      367-376.

30. Moser, G.C. and H. Meiss. 1977. A cytological procedure to screen mammalian cell mutants for cell-cycle related defects. Somat. Cell Genet. 3: 449-456.

31. Sheinin, R., Dubsky, M., Naismith, L. and J. Sigouin. 1984. Cell cycle arrest of heat-inactivated **ts** 2 mouse fibroblasts, temperature-sensitive for DNA synthesis. Manuscript in preparation.

32. Burstin, S.J., Meiss, H.K. and C. Basilico. 1974. A temperature-sensitive cell cycle mutant of the BHK line. J. Cell Physiol. **84**: 397-408.

33. Rossini, M., Baserga, S., Huang, C.H., Ingles, C.J. and R. Baserga. 1980. Defective RNA polymerase II in a cell cycle-specific temperature-sensitive mutant of hamster cells. J. Cell Physiol. **103**: 97-103.

34. Ingles, C.J. and M. Shales. 1982. DNA mediated transfer of an RNA polymerase II gene: Reversion of the temperature-sensitive hamster cell cycle mutant **ts** AF8 by mammalian DNA. Molec. Cell Biol. **2**: 666-673.

35. Robbins, E. and M. Scharff. 1966. Some macromolecular characteristics of synchronized HeLa cells. In **CELL SYNCHRONY**. (I.L. Cameron and G.M. Padilla) Academic Press, New York. pp. 353-374.

36. Sparkuhl, J. and R. Sheinin. 1980. Protein synthesis and degradation during expression of the temperature-sensitive defect in **ts** A1S9 cells. J. Cell Physiol. **105**: 247-258.

37. Sheinin, R., Darragh, P. and M. Dubsky. 1978. Some properties of chromatin synthesized by mouse cells temperature-sensitive in DNA replication. J. Biol. Chem. **253**: 922-926.

38. Sheinin, R. and P.N. Lewis. 1980. DNA and histone synthesis in mouse cells which exhibit temperature-sensitive DNA synthesis. Somat. Cell Genet. **2**: 227-241.

39. Sheinin, R., Humbert, J. and R. Pearlman R.E. 1978.
     Some aspects of eukaryotic DNA replication.
     Ann. Rev. Biochem. **47**: 331-370.

40. Sheinin, R. Setterfield, G., Dardick, I., Kiss,
     G. and M. Dubsky. 1980. Relationships between
     chromatin structure and replication in mouse
     L-cells. Canad. J. Biochem. **58**: 1359-1369.

41. Setterfield, G., Sheinin, R., Dardick, I., Kiss,
     G. and M. Dubsky. 1978. Structure of interphase
     nuclei in relation to the cell cycle. J. Cell
     Biol. **77**: 246-263.

42. Colwill, R.W. and R. Sheinin. 1982. Novobiocin-
     sensitivity and chromatin conformation in wild-type
     and temperature-sensitive mouse L-cells. Canad.
     J. Biochem. **58**: 195-203.

43. Aubin, J.E. 1981. Immunofluorescence studies
     of cytoskeletal proteins. In **MITOSIS/CYTOKINESIS**.
     (Ed. A.M. Zimmerman and A. Forer) Cell Biology
     Series, Academic Press. New York. pp. 211-244.

METABOLIC AND KINETIC COMPARTMENTS OF THE CELL CYCLE

DISTINGUISHED BY MULTIPARAMETER FLOW CYTOMETRY

Zbigniew Darzynkiewicz

Memorial Sloan-Kettering Cancer Center

New York, New York

INTRODUCTION

All cellular constituents double in content during the cell cycle. With few exceptions (DNA, histones, other nuclear proteins or enzymes associated with DNA replication) their synthesis proceeds continuously, although often at varying rates throughout interphase. Therefore, at a given point of the cell cycle, the content of a particular constituent is a reflection of its rate of synthesis (prior to this point) and turnover. Accumulation of the product is also influenced by the rate of cell division inasmuch as during cytokinesis all constituents are divided between daughter cells and thus their content per cell is reduced two-fold. When cells undergo differentiation, unbalanced growth or enter quiescence, the respective rates of accumulation of particular products additionally change in relation to DNA replication (content). The association between cell growth measured by accumulation of some of these products (e.g. RNA or proteins) and cell progression through the DNA division cycle has been a subject of extensive investigation (reviews, 1-3). Yet, the exact nature of this association, especially the cause-effect relationship between cell growth and initiation of DNA replication, is still unclear (e.g. see 4).

Multiparameter flow cytometry provides accurate estimates of relative quantities of various cell constituents (reviews, 5-8). Because the DNA content, i.e. the para-

meter which allows discrimination of G1 v̲s̲ S v̲s̲ G2 + M
cells, is usually one of the cell features measured, bi-or
multi-variate analysis of such results yields the cell distribution with respect to the second constituent correlated
with their position in the cell cycle. Thus, multiparameter flow cytometry is the method of choice to study
the relationship between cell growth and progression
through the cell cycle, especially when the heterogeneity
of the cell population has to be taken into account.
However, whereas the degree of advancement of the cell
through S can be accurately estimated based on the DNA
content, the cells' age distribution in G1 or G2 + M are
generally unknown and can be only inferred from the second
variable.   Also, in most measurements, discrimination
between G2 and M cells is not possible.

     Despite   these   shortcomings,   the   multiparameter
analysis of a particular cell population providing a
"snapshot" or static measurement of the cell cycle distribution, offers a wealth of information on the relationship
between the cell's metabolic state and its position in the
cell cycle. During the past decade, we have analysed a
variety of cell types, both normal and of tumor origin,
using this approach. In addition to cells growing exponentially, quiescent cells undergoing transition to the
proliferative compartment, differentiating cells and cells
in unbalanced growth were investigated.   Based on these
studies it was possible to recognize certain characteristic
features of cell populations common to different cell types
and to observe associations between these features and the
kinetic behavior of cells. We proposed that these features
could serve as metabolic markers distinguishing distinct
subcompartments located within the traditional phases of
the cell cycle (9-11). The presence of these compartments
in still other cell systems was confirmed by numerous
authors using our or other techniques (12-26).

     The combination of the multiparameter flow cytometry
with kinetic measurements makes it possible to correlate
the "snapshot" observations with the rate of cell progression through the cycle.   Three different experimental
designs are generally used. The first design is based on
studies of synchronized, often arrested populations.
Following stimulation or release from the arrest, their
progression is measured by sequential "snapshots"; by
analysis of changes in DNA content it is possible to

estimate the rates of cell entrance into S, S traverse or entrance into G2. Such techniques were used to measure cell kinetics in relation to RNA content in cultures of stimulated lymphocytes (27) or in cultures of CHO cells (28).

The second design makes use of the thymidine analog, 5-bromodeoxyuridine (BrdUrd). This analog, when incorporated into DNA during S phase, changes the stainability of DNA with several fluorochromes (29). The presence of the incorporated BrdUrd can also be detected immunochemically (30). Cells which undergo DNA replication in the presence of BrdUrd can thus be quantitated making it possible to study their rate of entrance into S, or to discriminate cycling from noncycling cells. Multiparameter analysis of cells which incorporate BrdUrd can be obtained by combination of Hoechst 33358 and propidium iodide (PI) to stain BrdUrd substituted-vs total-DNA (31), by simultaneous staining of DNA vs RNA with acridine orange (AO) (32) or by differential staining of native vs denatured DNA by AO to enumerate cells in mitosis (33).   Recently, Dolbeare et al (34) developed a method of simultaneous analysis of DNA content (PI-staining) and BrdUrd incorporation based on use of anti-BrdUrd antibodies. Because of its high sensitivity, the latter technique offers a powerful tool to analyse cell kinetics both in vitro and in vivo.

The third experimental design is based on the principle of stathmokinesis. We have proposed this design to study the rates of cell traverse through several points of the cell cycle simultaneously and have applied it to exponentially growing and drug perturbed cultures (35). The technique has been further refined (36) and used extensively in studies of drug effects on the cell cycle (37,38).

The present review updates these techniques and provides the most recent illustrations of the application of these methods to cell cycle analysis. The relationship between cell growth, as measured by the accumulated products such as RNA or proteins, and cell progression through the division cycle is discussed.

STATIC MEASUREMENTS

RNA and Protein Content

Fig. 1 illustrates most typical distributions of individual cells with respect to their DNA vs RNA or DNA vs

protein content when two parameters are measured simultaneously. This type of distribution is typical of any
exponentially growing population regardless of the cell
system (9,10,39). The DNA vs RNA and DNA vs protein cytograms in Fig. 1 are strikingly similar to each other
despite the fact that DNA measurements are more accurate
after staining with Hoechst 33342 than with AO, resulting
in better resolution of G1 vs S vs G2 + M populations. The
characteristic feature of these distributions is the
presence of the threshold RNA (marked by arrow) or protein
content in G1. As is evident, cells with a subthreshold
content of these constituents (G1A) do not enter S phase.
Thus, in the exponentially growing, unperturbed population,
cells in G1 have to accumulate the threshold content of RNA
and protein prior to entrance into S, and it is clear that
this restriction is rigorously controlled. In the case of
DNA-RNA distributions, we have studied at least 40
different cell types both normal and of tumor origin. All
these cell systems were characterized by this threshold.
Because the G1 cell distributions with respect to RNA or
protein are generally unimodal (9,10,39), the G1A to G1B

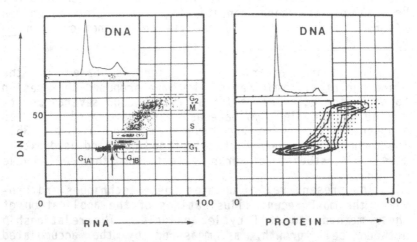

Fig. 1: The cytogram and contour map representing the
distribution of CHO cells from exponentially growing cultures with respect to their DNA vs RNA and DNA vs protein
values, respectively. Cells were stained with AO for
DNA vs RNA analysis and with Hoechst 33342 and Rhodamine
640 for DNA vs protein measurements.

transition appears to be continuous. To have an objective approach to discriminate between G1A and G1B cells, a gating window can be located at the early one-third of S phase (based on the DNA content) and the cells measured within the window to obtain the mean value and standard deviation with respect to RNA (or protein) content. Then the threshold is established at the level of the mean value minus three standard deviations of these early-S cells, as shown by the arrow (Fig. 1). Thus, by definition, the G1A cells have a RNA (or protein) content significantly lower than the early-S cells whereas the G1B cells are similar to the early-S cells. As we have emphasized in earlier publications (9-11) and as will be discussed further, the kinetic properties of the G1A and G1B populations appear to be different.

A technical point should be stressed here which is of importance in discriminating the G1A and G1B populations. Namely, the presence of any dead, or broken cells in cultures precludes detection of the threshold because the broken S phase cells, depleted of all or a portion of the cytoplasm will be located on the cytograms below the threshold. The dead cells, however, can be removed by preincubation of cultures with trypsin and/or DNase I prior to fixation and staining (10,38). Detection of the threshold also requires good accuracy (resolution) of the measurements. When the c.v. values of the mean DNA content of G1 are higher than 6-8%, the distinction between G1A and G1B becomes obscured unless the G1A population is very prominent (G1A-arrest).

The method for simultaneous, direct determination and correlation of DNA, RNA and protein in individual cells was recently developed based on use of the three-laser flow cytometer (40). In addition to these three parameters, the RNA/DNA and RNA/protein ratios can also be measured and recorded in the list mode so that each of these five measurements can be correlated with any other one for a given cell. An example of such measurements of CHO cells is shown in Fig. 2. The cells in this experiment were just entering the sub-plateau phase of cell growth, hence there are fewer S, G2 and M cells and more G1 cells with low RNA and protein content in comparison with cells shown in Fig. 1. Analysis of this multiparametric measurement yields many details on the relationship between cell growth and position in the cell cycle.

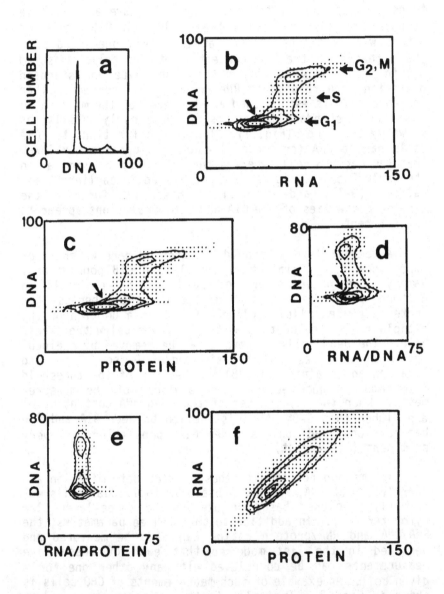

<u>Fig. 2:</u> Simultaneous measurements of DNA, RNA and protein content as well as RNA/DNA and RNA/protein ratios of CHO cells. The cells were stained with Hoechst 33342, pyronin Y and FITC and their fluorescence measured in a three-laser flow cytometer (40).

The DNA vs RNA and DNA vs protein cell distributions in Fig. 2 (b,c) are similar to each other and the cytograms shown in Fig. 1 despite the entirely different methods of cell staining used in the experiments. Judging from the width of the G1, S or G2 + M clusters and from the standard deviation of the mean values (not shown) the populations are more heterogeneous when an analysis is based on protein rather than RNA content. This is also evident in Fig. 1 and confirms our earlier observations (11). The aforementioned thresholds of RNA and protein content in G1 are also clearly manifested (arrows).

In this multiparameter-correlated analysis (Fig. 2) it is possible to estimate whether the low RNA G1 cells (G1A) also have low protein content, i.e. whether the G1A populations discriminated either by RNA or protein content are the same. Indeed, by gating the subthreshold G1 cells based on RNA content and re-plotting the gated population with respect to DNA vs protein, we observe that the subthreshold (G1A) populations discriminated by RNA and protein content are nearly identical (not shown).

Analysis of the cellular RNA/DNA ratio (Fig. 2d) in relation to DNA content reveals a characteristic pattern reflecting changing rates of DNA replication and transcription during the cell cycle. Thus, during G1 when DNA content is constant, cells accumulate increasing quantity of RNA which results in the G1 phase being heterogeneous with respect to RNA/DNA ratios. A critical RNA/DNA ratio (arrow) reflects the RNA threshold discussed above. During progression through S phase the rate of DNA replication exceeds RNA accumulation giving rise to the slanted negatively skewed slope of the S-cluster. Cells in G2 + M have an RNA/DNA ratio similar to that of the majority (peak values) of the G1 cells.

The ratio of RNA/protein remains remarkably constant and uniform throughout the cell cycle (Fig. 2e) with only a few G1 cells having elevated ratios contributing to the skewed distribution of the G1 population. The RNA and protein content of individual cells are highly correlated (Fig. 2f). The multiparameter analysis, as shown in Fig. 2, is of great value in estimating unbalanced growth, e.g. as induced by various antitumor drugs.

Changes in the DNA and RNA distribution of cells undergoing transition to quiescence are shown in Fig. 3. The upper panels (a-c) illustrate changes in 3T3 cells subjected to growth at low serum concentration or in confluent cultures. Exponentially growing 3T3 cells (a) show a well defined RNA threshold in G1 which allows the discrimination of G1A and G1B subpopulations, as discussed above for CHO cells. When cells are maintained at 0.5% serum for 72 hr (b) a marked suppression in the number of S and G2 + M cells is apparent. Most cells are arrested in G1; the arrested cells are characterized by low RNA values, typical of the G1A population. Thus, by the criterium of RNA content, those cells are arrested in G1A. Following addition of serum, most of those G1A-arrested cells enter S phase after about a 12 hr delay. Prolonged incubation of cells in serum deprived media results in loss of cell viability without further decrease in RNA content. 3T3 cells, however, can enter a deeper state of quiescence when maintained at confluence for an extended period of time.

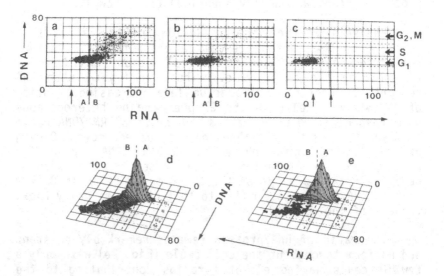

Fig. 3: RNA and DNA values of 3T3 cells in (a) exponential growth, (b) in cultures with 0.5% serum and (c) in confluent cultures. The histograms show RNA and DNA values of L1210 cells during exponential growth (d) and following treatment with 1mM sodium butyrate (e).

Namely, as shown in Fig. 3 panel c, cell growth for an additional 4 days after having already reached confluence, results in their arrest in G1; the RNA content of these arrested cells, however, is now below the level of the G1A population of the exponentially growing, or G1A-arrested cells. These cells are viable and when trypsinized and replated at lower density resume progression through S after about a 16 hr delay. Such deeply quiescent G1 cells, characterized by markedly lowered RNA content, can thus be distinguished as a separate category, (G1Q cells) having distinct metabolic and kinetic properties.

The lower panels (d and c) in Fig. 3 provide an example of cell arrest in the G1A compartment. This data presents L1210-leukemia cells growing exponentially (d) and in cultures containing 1mM n-butyrate (e). It is quite clear that in the presence of n-butyrate the number of cells in S, G2 and M is diminished and most G1 cells have an RNA content typical of the G1A population. Extended growth of leukemic cells in the presence of n-butyrate results in loss of their viability rather than further decrease of the RNA content to the level of G1Q cells.

Another example of G1Q cells are nonstimulated peripheral blood lymphocytes. These cells have minimal RNA content and upon stimulation require at least 24 hr to enter S phase. Fig. 4 illustrates DNA and RNA values of lymphocytes stimulated to proliferation in cultures. The upper panel (a) shows cells in 6-day old cultures containing mixed lymphocytes from two unrelated donors ("mixed allogeneic reaction"). In such a culture, there are unstimulated (nonresponding) G1Q cells as well as the cells undergoing proliferation present. The nonresponding cells have low RNA content identical to nonstimulated cells prior to culturing. Responding cells triggered to proliferate by the antigen presented by the allogeneic lymphocytes have an RNA content several times higher than G1Q cells. Their number is substantially increased after 6 days in culture, due to several rounds of cell division. At that time, there are few cells in transition between the G1Q and the G1A,B compartments and, therefore, the G1Q peak is totally separated from the cycling population.

The cytogram representing stimulation of lymphocytes in cultures containing the mitogen phytohemaglutinin (PHA) is shown in Fig. 4b. Nonstimulated (nonresponding) G1Q

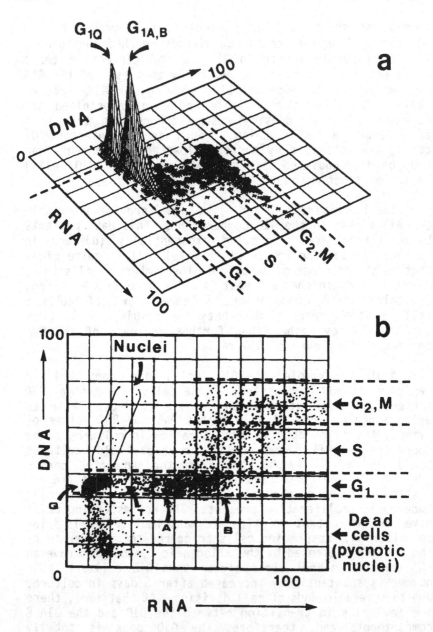

Fig. 4: RNA and DNA values of human peripheral blood lymphocytes stimulated to proliferation in allogeneic mixed cell cultures (a) and in PHA-treated cultures (b).

cells have very low RNA. The mitogen-responsive lymphocytes depending on DNA content can be classified as G1, S
or G2 + M. Among G1 cells, cells in G1A and G1B can be
recognized as in Fig. 1. Few cells in transition (G1T) with
RNA values intermediate between G1Q and G1A are seen in
such cultures; these cells predominate, however, during the
first day of stimulation with PHA. Dead cells with
pyknotic nuclei have lowered DNA stainability. Broken
cells and isolated nuclei have low RNA but normal DNA
content and may be erroneously taken as quiescent cells.
As described before (10,38) pretreatment with DNase I and
trypsin removes damaged cells and nuclei and thus permits
discrimination between damaged cells and intact cells with
low RNA content.

Recognition of G1Q cells is also possible based on RNA
content of isolated nuclei. Fig. 5 shows DNA and RNA values
of nuclei isolated from the liver of a 14 week-old mouse.
Two populations of G1-nuclei are present. It is evident

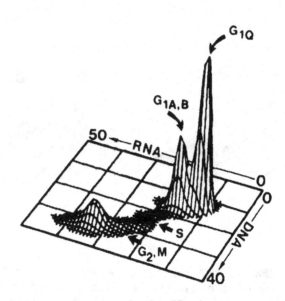

Fig. 5: RNA and DNA content of nuclei isolated from liver
cells of a 14-week-old mouse.

that cells entering S phase originate from the high nuclear RNA population. In older mice, when proliferation of hepatocytes is minimal, nearly all diploid nuclei belong to the low-RNA (G1Q) population, whereas stimulation of hepatocyte proliferation in regenerating liver coincides with disappearance of the G1Q population (41). Studies of nuclear rather than total cell RNA content offers certain advantages inasmuch as isolation of nuclei is often a simpler task (e.g. from such tissues as liver or certain tumors) than obtaining a preparation of well dispersed intact cells suitable for flow cytometry. Furthermore, nuclear RNA may be a more sensitive marker of changes in genome transcription and may react more rapidly to any metabolic change which affects transcription.

Although in the majority of the cases we observed that cells can be arrested in either G1A or G1Q, there were situations in which quiescent cells were characterized by an S- or G2-DNA content (10). The term SQ and G2Q was proposed to characterize such cells (10). The presence of

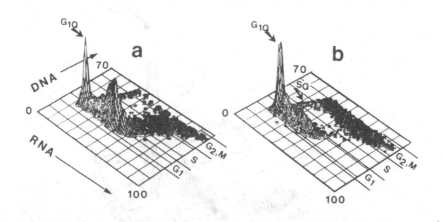

Fig. 6: RNA and DNA content of mononuclear cells from the bone marrow of the patient with chronic myeloid leukemia during blastic crisis prior (a) and 24 hr after (b) treatment with desacetylvinblastine amide sulfate. There were more than 85% blast cells in these preparations as evaluated by cell morphology.

SO and G2Q cells was recently confirmed in a number of laboratories (e.g. 12). An example of kinetically inactive cells with a DNA content typical of S or G2 cells is shown in Fig. 6. This figure represents histograms of blast cells from the blood of a patient with chronic myeloid leukemia in blastic crisis. Panel (a) shows cells obtained prior to treatment and indicates the presence of two subpopulations, one with low RNA content and the other one with high RNA content. Both subpopulations have prominent G1 peaks and both have cells with an S- and G2 DNA content. Panel (b) shows cells from the same patient obtained 24 hr after treatment with the vinblastine analogue desacetyl-vinblastine amide sulfate, which arrests cells in mitosis. It is quite clear that the treatment perturbed only the high-RNA subpopulation. In this subpopulation, there is an increase in number of cells in G2 + M and a proportional decrease of cells in G1 and S. In contrast, the low-RNA subpopulation appears to be unchanged; it still contains a distinct G1 peak and cells with an S DNA content. Although not direct proof, this example provides strong, indirect evidence that the low-RNA population consists of G1Q, SO and G2Q cells, resistant to cell cycle specific drugs. The presence of such cells was observed in several other cases of chronic myeloid leukemia during blastic crisis.

Mitochondrial Activity

There are several cationic fluorescent probes which show high affinity towards mitochondria of living cells. Rhodamine 123 (R123) is the most commonly used probe of mitochondria. Uptake of this fluorochrome depends both on number (mass) of mitochondria per cell and the electronegativity of the mitochondrial membrane. We have studied the uptake of R123 in relation to cell position in the cell cycle, quiescence and differentiation (43). An example of changes in R123 uptake during the transition of quiescent cells to the cell cycle is illustrated in Fig. 7. In this figure the second parameter is forward light scatter, which is closely related to cell size. Nonstimulated lymphocytes (G1Q) are very uniform, having low light scatter and fluorescence values. The population of PHA-stimulated lymphocytes consists of responding (cycling) and non-responding (G1Q) subpopulations. Discrimination of these subpopulations is possible based on both R123 fluorescence and light scatter; the fluorescence, however, appears to offer better discrimination.

Friend erythroleukemia cells induced to differentiate in the presence of DMSO exhibit markedly diminished uptake of R123 in comparison with their exponentially growing counterparts (43). Also, L1210, CHO and Friend erythroleukemia cells in stationary cultures bind approximately twice less R123 than their counterparts in exponentially growing cultures. Because of high intercellular variability in the binding of R123 in most cell types, the discriminatory power of this fluorochrome is lower than, for example, of the RNA content.

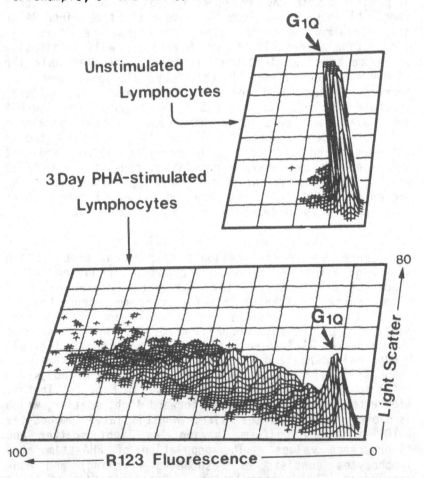

Fig. 7: Forward light scatter and green fluorescence of unstimulated and stimulated human lymphocytes after staining with the mitochondrial probe R123.

Chromatin Changes

Profound changes occur in both the gross- and molecular-structure of nuclear chromatin during the cell cycle as well as during cell transition to quiescence or differentiation (review, 3). We have developed a technique to study chromatin changes based on differences in sensitivity of DNA in situ to denaturation (44,45). The factors responsible for the variation in in situ DNA sensitivity to denaturation are poorly understood. It is believed that histones and perhaps other nuclear proteins, by providing local counterions contribute to the stability of DNA in situ, and that their postsynthetic modifications alter DNA stability (3). The technique used to study DNA denaturability in intact cells is based on subjecting RNase treated cells to heat, or acid and subsequent staining with the metachromatic fluorochrome, acridine orange. After partial denaturation of DNA by heat or acid, acridine orange stains the nondenatured DNA sections green whereas interactions of the dye with the denatured sections result in red luminescence (46). Thus, the relative proportions of red- and green-luminescence of so stained cells represent portions of the denatured- and native- DNA, respectively.

In general, in the majority of cell types, the sensitivity of DNA to denaturation correlates with the degree of chromatin condensation. The DNA most sensitive to denaturation is in mitotic cells as well as in quiescent cells characterized by the condensed chromatin (G1Q). The most resistant is DNA in late G1 (G1B) and early-S cells. This pattern of staining is shown in Fig. 8. Quiescent, non-stimulated lymphocytes (a) are very uniform having, after treatment at pH 1.5, approximately equal proportions of green and red luminescence. Lymphocytes stimulated for 18 hr with PHA (b) consist mostly of cells undergoing transition to the cell cycle (G1T) prior to entrance to S. These cells have a wider distribution with the major sub-population characterized by increased green- and lowered red-luminescence in comparison with G1Q cells. Sub-populations of cells in G1Q, G1A,B, S, G2 and M can be distinguished in lymphocyte cultures stimulated by PHA for 3 days (c,d). The number of cells in M is increased, concomitant with a decrease in the proportion of G1 cells in the culture treated additionally with Colcemid (d). The peaks and ridges, prescribed to particular subphases as

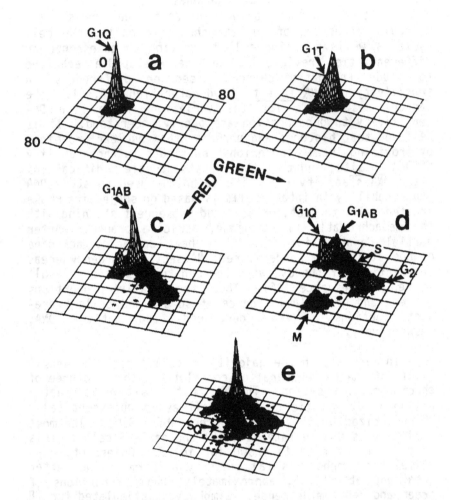

Fig. 8: Sensitivity of DNA in situ to acid-induced denaturation during stimulation of human lymphocytes (a-d) and in
mononuclear cells from bone marrow of a patient with
chronic myeloid leukemia during blastic crisis (e). Nonstimulated (a) and PHA stimulated lymphocytes 18 hr (b) and
3 days (c,d) after PHA. The last culture (d) was
additionally treated for 8 hr with Colcemid, to arrest
cells in mitosis prior to harvesting. After staining with
AO, the proportions of nondenatured- and denatured-DNA are
represented by green- and red-fluorescence, respectively
(44).

marked on the histograms (a-d) were identified by studies of synchronized populations (44,45).

Fig. 8e illustrates leukemic cells from the patient with chronic myeloid leukemia during blastic crisis. Two subpopulations of S phase cells can be discerned in this preparation. The subpopulation of S-cells with DNA almost as sensitive to denaturation as G1Q cells most likely represents SO cells. This subpopulation is not present when all cells are progressing through the cell cycle in short term cultures (47). Also, proportions of S-cells characterized by high DNA sensitivity to denaturation correlate with S-phase cells having low RNA content, as in Fig. 5.

The pattern of cell stainability as shown in Fig. 8 with the possibility of discriminating G1Q, G1A,B, S, G2 and M cells can be almost duplicated by simultaneous staining of nuclear proteins vs DNA (48). A threshold protein content in G1 is apparent with distribution of cells in G1A and G1B as based on DNA sensitivity to denaturation. Also, cells in mitosis have little proteins associated with the chromosomes and thus a distinction between G2 and M can be made (48). Based on this similarity, it is tempting to speculate that the nonhistone proteins, the content of which increases during G1 and which may dissociate from chromosomes during mitosis, may modulate the stability of DNA in situ and be responsible for the observed patterns (Fig. 8). In contrast, the histone/DNA ratio remains rather constant throughout the cell cycle.

KINETIC STUDIES

Sequential "Snapshots" of Synchronized Populations

An example of an analysis of cell cycle progression using sequential measurements of synchronized cells is given in Fig. 9. In this experiment, peripheral blood lymphocytes were stimulated by PHA and hydroxyurea was added 24 hr following PHA for the next 24 hr to prevent progression of cells through S phase. Thus, stimulated lymphocytes which otherwise asynchronously enter S phase, were held at the G1/S interphase for up to 24 hr by hydroxyurea. The cells were released from the hydroxyurea block at 0 time and their progression through S studied during the subsequent 8 hr. Three cell subpopulations can be

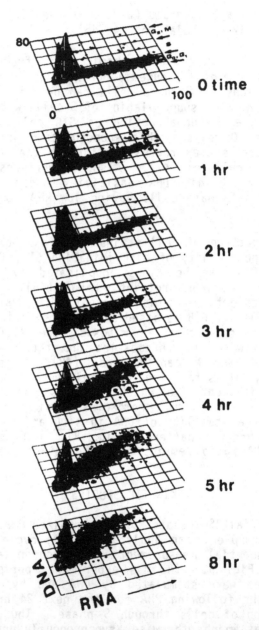

Fig. 9: Progression through S phase of lymphocytes stimulated by PHA and synchronized at the G1/S boundary, in relation to their RNA content. Synchronized cultures were released from the block and cellular RNA and DNA was measured 0,1,2,3,4,5 and 8 hr after the release.

recognized on the histogram at 0 time. The minor peak at
lowest RNA (10 channels) represents cells in G0 (G1Q)
which did not respond to PHA. The major peak (20 channels)
represents cells in transition from G1Q to the cell cycle;
such cells predominate during the initial 48 hr of stimulation. The cells with highest RNA, represented by the
ridge (30-100 channels) are already in the cell cycle but
arrested by hydroxyurea. Upon removal of the blocker, the
cells traverse S phase. Their rate of progression is
highly correlated with their RNA content. Whereas, for
instance, high RNA cells have already reached G2 by 4 hr,
the low RNA cells traversed only a small portion of S. By
gating analysis it is possible to calculate the rates of S-
phase-traverse for various fractions of S-phase cells
depending on their RNA content (27).

The analysis as demonstrated in Fig. 9 can be applied
to a variety of synchronized populations, e.g. in studies
of rates at which G1Q cells enter the cell cycle (49) or how
fast post-mitotic cells progress through G1 and S depending
upon RNA content they inherit during cytokinesis (28).

BrdUrd Incorporation

BrdUrd detection by flow cytometry offers an alternative to ^{3}H-Tdr autoradiography. The advantages of this
technique, which is already in wide use, are obvious to
anyone familiar with laborious procedures of autoradiography. The scope of this article does not allow a review
of the technical approaches for BrdUrd detection or their
applications. An example, however, is given to illustrate
the principle of the technique and the character of the
data (Fig. 10).

The upper three panels (a,b,c) in Fig. 10 show RNA/DNA
distribution of PHA-stimulated lymphocytes grown in the
absence of BrdUrd on the 4th (a), 6th (b) and 8th (c) day
after addition of PHA. As is evident, the cells show
maximal stimulation on day 4 and then return to quiescence;
a decrease in RNA content concomitant with decreased
proportion of S-G2-M cells characterizes this process. The
three lower panels illustrate parallel cultures which were
exposed to BrdUrd for approximately one cell generation;
(d) is a culture parallel to (a) but BrdUrd was present
during 20 hr prior to harvesting; (e) is the culture
parallel to (b) also treated with BrdUrd for 24 hr prior to
harvesting, i.e. on the 6th day, whereas (f) is parallel to

(c), i.e., exposed to BrdUrd for 20 hr during the 6th day
and harvested on day 8. Thus, during cell proliferation
the cultures were exposed to the precursor which resulted
in 40% lower DNA stainability of cycling cells when stained
(32) with AO.

When cells return to quiescence and their RNA content
decreases, the previously cycling population can still be
recognized (f) despite the fact that the DNA content of
those cells is now not much different from that of the
nonproliferating cells. Thus, the mitogen-responsive
("memory") cells can be distinguished from the nonresponding ones even after the cells enter quiescence and
can no longer be discriminated by metabolic parameters such
as RNA or protein content.

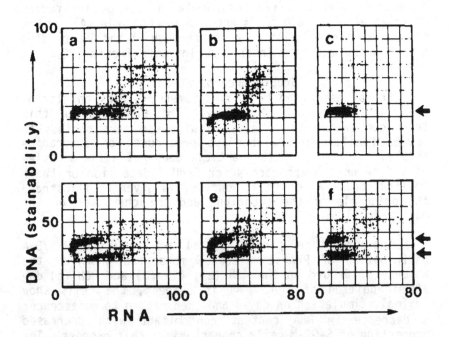

Fig. 10: RNA and DNA stainability with AO of PHA stimulated
lymphocytes grown in the absence (a-c) and presence of
BrdUrd (d-f). The arrows mark the G1 clusters.

As illustrated in Fig. 10, the BrdUrd technique can be used to: 1) recognize the cycling cell population during proliferation; 2) to pre-label the cycling cells for their later identification when they enter quiescence; and 3) correlate metabolic parameters (e.g. as RNA content in Fig. 10) with proliferation.

A recently introduced technique which combines DNA staining with BrdUrd detection by monoclonal antibodies (34) is the most sensitive technique available to recognize DNA replicating cells. It is, therefore, expected it will find wide application, especially in _in_ _vivo_ cytokinetic studies.

Stathmokinetic Studies

The principle of the stathmokinetic experiment to analyse cell kinetics such as proposed by Puck and Steffen (50) can be combined with multiparameter flow cytometry (35-38). Such a combination offers a plethora of information on rates of progression through various portions of the cell cycle and can be applied both to nonperturbed and drug-perturbed cultures. In this review, a single example is given of the application of this method to the analysis of cell progression through G1 and cell exit from G1 (Fig. 11). This example is pertinent to further discussion on the role of the G1A compartment in the cell cycle.

The three panels in Fig. 11 show exponentially growing CHO cells, non-treated (a) and treated for 2 (b) or 4 hr (c) with Colcemid. This type of staining, based on changes in both DNA content and chromatin conformation allows for the discrimination of the G1A and G1B compartments in G1 as well as the discrimination between G2 and M cells (44,45). As discussed before, the distinction is based on differences in sensitivity of DNA to denaturation, represented as α_t (35). The entrance to S phase occurs when cells attain the threshold α_t; thus, the subthreshold G1A subpopulation can be distinguished as in the case of RNA or protein content (Fig. 1). The G1B cells have DNA sensitivity similar to that of early S cells.

When the stathmokinetic agent such as Colcemid is added to exponentially growing cultures, the kinetics of cell exit from G1, transition from G1A to G1B, and entrance to M can be measured as illustrated in Fig. 11. Namely, by establishing the gating thresholds for the G1A, G1 and M

populations, the number of cells in each population at
different times after addition of the stathmokinetic agent
can be plotted. The plot (d) represents only changes in the

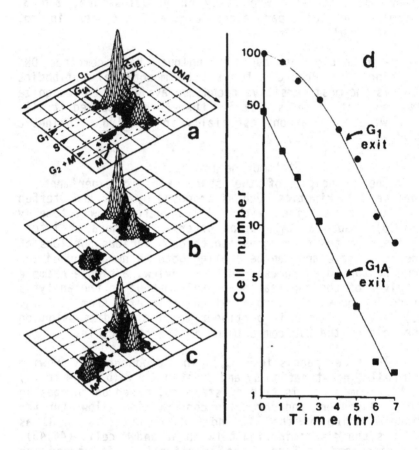

Fig. 11: Kinetic experiment showing measurement of the
rate of cell exit from G1 and G1A. Exponentially growing
cultures of CHO cells were treated with Colcemid and the
cells were sampled at 0 (a), 2 (b) and 4 hr (c) during
stathmokinesis. After partial denaturation of DNA in situ
and staining with AO, the subpopulations as shown in (a)
can be distinguished. The number of cells remaining in G1A
and G1 is then plotted vs the time after addition of the
stathmokinetic agent.

G1A and G1 compartments. The main feature of this data is that the exit from G1 or G1A has an exponential component manifested as a straight slope on a semi-logarithmic scale (35,36). In the case of G1A population, this exponential slope is evident from the onset of stathmokinesis. This data reproduced numerous times on different cell types (35-39), demonstrate that cells have very heterogeneous residence times in G1A with a characteristic exponential or "probablistic-like" component. More extensive analysis of the stathmokinetic experiment is provided elsewhere (51).

METABOLIC SUBCOMPARTMENTS OF THE CELL CYCLE

Examples of cell subpopulations showing different metabolic properties which can be distinguished by flow cytometry were illustrated in the first portion of this Chapter. Based on these differences, we have proposed to subdivide the cell cycle into several distinct subcompartments (9,10). Such subdivision offers higher accuracy of cell classification in comparison with traditional distinction of the four main phases of the cell cycle. A summary of characteristics and typical examples of these compartments are given below.

1) G1A compartment is characteristic of exponentially growing populations. It is comprised predominantly of early G1, post-mitotic cells which have an RNA or protein content significantly lower than cells in early S-phase. The chromatin of G1A cells is also distinct from the chromatin of early-S cells; the difference is manifested as increased sensitivity of DNA in situ to acid-induced denaturation. The cell residence times in G1A have a characteristic, exponential component which is responsible for heterogeneity of transit times through G1 or perhaps for heterogeneity of cell generation times in general. Assuming the Mitchison's concept of the cell cycle (2), it was postulated (11) that the G1A phase represents a gap in the DNA division cycle. During this phase the cells are in a growth cycle accumulating certain metabolic constituents up to a critical threshold. Thus, G1A appears to be the "growth" or "equalization" subphase. Cells' residence times in G1A are inversely proportional to the amount of the metabolic constituents that they inherit during cyto-kinesis.

2) G1A-arrested cells have metabolic characteristics of G1A cells. In contrast to the latter, however, they do not progress towards S. Cells maintained in cultures deprived of serum (e.g. 3T3 cells) or growing in the presence of sodium butyrate (L1210) show characteristics of G1A cells. Upon release from the arrest, a delay of between 10-12 hr is observed prior to their entrance to S.

3) G1B cells are present in exponentially growing populations. The RNA or protein content of G1B cells is similar to that of early-S cells. Likewise, the chromatin structure of G1B cells reflected by the sensitivity of their DNA in situ to denaturation resembles that of early-S cells. Most likely, G1B cells are committed to enter S phase and are in the preparatory phase of DNA replication. G1B and early-S cells exhibit the highest resistance to DNA denaturation in situ.

4) Traditional S phase comprises cells replicating DNA. In unperturbed, exponentially growing populations a threshold RNA or protein content is required prior to entrance to S. However, the cells may bypass this require- ment and under certain experimental conditions may enter S with a "subthreshold" quantity of RNA or protein (4). A progressive increase in DNA sensitivity to denaturation occurs during S.

5) Cells in G2 accumulate additional quantities of RNA and protein and initiate chromatin changes preparatory to mitosis. The sensitivity of DNA in situ in G2 cells to denaturation is similar to that of G1A cells and is clearly higher than that of the S or G1B cells.

6) Cells in mitosis (late prophase, anaphase, meta- phase, telophase) can be recognized as having the most condensed chromatin, which coincides with the highest degree of sensitivity of DNA in situ to denaturation. Mitosis is unequal, producing two daughter cells with dif- fering quantities of RNA or proteins and thus generating heterogeneity in the cell cycle (11). Unequal mitosis may, to a large extent, contribute to the heterogeneity and exponential-like component in the transit times of cells through G1A. A mathematical model of the cell cycle based on the unequal distribution of metabolic constituents during mitosis has recently been proposed by us (52).

7) G1Q cells are characterized by minimal metabolic activity. They have low RNA and protein content, very condensed chromatin, high sensitivity of DNA in situ to acid- or heat-induced denaturation and a low number of mitochondria. By all these parameters, G1Q cells are significantly different from their exponentially growing counterparts. Upon stimulation, these cells require a long (16-24 hr) prereplicative phase during which they synthesize RNA, proteins, develop new mitochondria, and decondense chromatin. Nonstimulated peripheral blood lymphocytes or 3T3 cells maintained at confluency for an extented time are examples of G1Q cells.

8) SQ cells are characterized by an S-DNA but low RNA content and have DNA more sensitive to denaturation than cycling S cells. SQ cells, however, either do not progress through the cell cycle or do so very slowly. Cells with these characteristics were observed in human leukemia cell lines co-incubated with activated macrophages (53) and in the bone marrow of patients with chronic myeloid leukemia during blastic crisis (47).

9) G2Q cells share characteristics of SQ cells except they have a G2 DNA content.

10) S-arrested cells have all the metabolic features of S-phase cells except they do not replicate DNA. In certain tumors, anoxic cells exhibit these characteristics (54). Cells arrested in S by most of the anti-tumor drugs (e.g. araC, hydroxyurea) develop, with time, signs of unbalanced growth. The degree of unbalanced growth can be estimated from the deviation in the ratio of RNA/DNA, or protein/DNA from the normal values. When cells are in negative unbalanced growth (e.g. after treatment with actinomycin D or cycloheximide) they resemble SQ cells.

11) G2-arrested cells are similar, by metabolic criteria to G2 cells. Many antitumor drugs arrest cells in G2. Such arrested cells often develop signs of unbalanced growth, as the S-arrested cells.

12) G1D cells are fully differentiated and have a G1 DNA content. Differentiated cell phenotypes may be charac- terized by low- (granulocytes, nucleated erythrocytes), moderate- (macrophages, hepatocytes) or high-(plasma

cells)-RNA content. Unlike G1Q, these cells cannot enter, the cell cycle, unless under unusual circumstances (cell fusion).

13) Cells in transition. When cells undergo transition, e.g. from the G1Q state to proliferation, for a certain period of time they possess intermediate metabolic features. Peripheral blood lymphocytes stimulated by mitogens exhibit characteristics of cells in transition from quiescence to proliferation (G1T cells) during the first 24 hr of stimulation. Depending on their DNA content during the transition these cells can be classified as G1T, ST or G2T (9).

CELL GROWTH AND DNA DIVISION CYCLE;
THE MULTIPARAMETER FLOW CYTOMETRY EVIDENCE

As mentioned before, simultaneous measurements of two or more parameters, one of which is DNA content, provides information on cell growth and cell progression through the DNA division cycle. The subject of the association between cell growth and DNA division, has been most clearly formulated by Mitchison (2), and has been the subject of extensive investigations during the past three decades. Yet, still many uncertainties remain. A recent review by Baserga (4) offers a short but excellent coverage of this subject. In this article Baserga proposes that cell growth and initiation of DNA replication are coordinated during exponential cell growth. Thus, there must be mechanisms which sense accumulation of RNA or perhaps total or nuclear protein and transfer the signal to the regulatory point of initiation of DNA replication. Cell growth in size and cell progression through the DNA division cycle, however, can be dissociated, e.g. by suppressors of protein synthesis (55), viruses (56) or other treatments. Exposure of cells to alkaline media was reported to trigger DNA replication in the absence of cell growth in size (57); this data, however, could not be reproduced (58). Still, there is enough experimental evidence indicating that cell growth and initiation of DNA synthesis can be decoupled to support Baserga's postulate that cell growth and the DNA division cycle may be under distinct regulatory mechanisms. The data obtained by multiparameter flow cytometry is in support of this contention.

The evidence provided by flow cytometry, however, indicates that during nonperturbed, expontential growth in cultures, the coupling of cell growth and DNA replication is very strong. Thus, in exponentially growing cell populations, when rigorous controls of cell viability are exercised fewer than 0.01% cells in Gl with very low RNA content (low-RNA half of the GlA subpopulation) are observed to enter S phase. Furthermore, the coupling is also evident during S phase inasmuch as cells "overloaded" with RNA traverse S phase at much faster rates than their counterparts with a normal RNA content (18,27,28). Whereas normal progression through S is associated with an increase in RNA content per cell, progression of the "overloaded" cells does not involve any further RNA accumulation (27,28). All these observations suggest that although the regulatory site(s) for cell growth in size (e.g. transcription of rRNA messages) and DNA replication may be distinct, those sites are controlled in unison during nonperturbed cell growth. Further research is needed to assess to what extent regulation of those sites is coordinated in vivo, in normal tissues or during tumor cell growth, when conditions are different than those in exponentially growing cultures. The inability of tumor cells to enter a state of deep quiescence characterized by low rRNA (59) may point out that coordination of these sites may not be functioning during neoplastic growth.

REFERENCES

1)   Baserga, B. Multiplication and Division in Mammalian Cells. Marcel Dekker, New York, 1976.
2)   Mitchison, J.M. The Biology of the Cell Cycle. The University Press, Cambridge, 1971.
3)   Darzynkiewicz, Z. Pharmac. Ther. 21:143-188, 1983.
4)   Baserga, R. Exp. Cell Res. 151:1-5, 1984.
5)   Melamed, M.R., M.L. Mendelsohn and P.F. Mullaney (eds) Flow Cytometry and Sorting. John Wiley and Sons, New York, 1979
6)   Kruth, H.S. Anal. Biochem. 125:225-242, 1982.
7)   Traganos, F. Cancer Investig. (in press)
8)   Barlogie, B., N.M. Raber, J. Schumann, T.S. Johnson, B. Drewinko, D.E. Swartzendruber, W. Gohde, M. Andreeff and E.J. Freireich. Cancer Res. 43:3982-3897, 1983.
9)   Darzynkiewicz, Z., T. Sharpless, L. Staiano-Coico and M.R. Melamed. Proc. Natl. Acad. Sci. USA 77:6696-6699, 1980.
10) Darzynkiewicz, Z., F. Traganos and M.R. Melamed. Cytometry 1:98-108, 1980.

11) Darzynkiewicz, Z., H. Crissman, F. Traganos and J. Steinkamp. J. Cell Physiol. 112:465-474, 1982.

12) Allison, D.C., J.M. Yuhas, R.F. Ridolpho, S.L. Anderson and L.S. Johnson. Cell Tissue Kinet. 16:237-243, 1983.

13) Antel, J.P.. J.F. Oger, E. Dropcho, D.P. Richman, H.H. Kuo and B.G.W. Arnason. Cell Immunol. 54:184-192, 1980.

14) Barlogie, B., A.M.M. Maddox, D.A. Johnston, M.N. Raber, B. Drewinko, M.J. Keating and E.J. Freireich. Blood Cells 9:35-55, 1983.

15) Bauer, K.D. and L.A. Dethlefsen. J. Cell Physiol. 108:99-112, 1981.

16) Bauer, K.D., P.C. Keng and R.M. Sutherland. Cancer Res. 42:72-78, 1982.

17) Creasey, A.A., J.C. Bartholomew and T.C. Merigan. Exp. Cell Res. 134:155-160, 1981.

18) Fujikawa-Yamamoto, K. J. Cell. Physiol. 112:60-66, 1982.

19) Kristensen, F., C. Walker, F. Bettens, F. Joucourt and A.L. DeWeck. Cell. Immunol. 74:140-149, 2982.

20) Marder, P. and J.R. Schmidtke. Immunopharm. 6:155-166, 1983.

21) Monroe, J.G. and J.C. Cambier. J. Immunol. 130:626-631, 1983.

22) Nusse, M. and H.J. Enger. Cell Tissue Kinet. 17:13-23, 1984.

23) Richman, P.D. J. Cell Biol. 85:459-465, 1980.

24) Shapiro, H. Cytometry 2:143-150, 1981.

25) Wallen, A.C., R. Higashikubo and L.A. Dethlefsen. Cell Tissue Kinet. 17:65-77, 1984.

26) Watson, J.W. and S.H. Chambers. Br. J. Cancer 38:592-600, 1977.

27) Darzynkiewicz, Z., D.P. Evenson, L. Staiano-Coico, T. Sharpless and M.R. Melamed. Proc. Natl. Acad. Sci. USA 76:358-362, 1979.

28) Darzynkiewicz, Z., D.P. Evenson, L. Staiano-Coico, T. Sharpless and M.R. Melamed. J. Cell Physiol. 100:425-438, 1979.

29) Latt, S.A. J. Cell Biol. 62:546-550, 1974.

30) Gratzner, H.G. and R.C. Leiff. Cytometry 1:385-389, 1981.

31) Bohmer, R.M. and J. Ellwart. Cell Tissue Kinet. 14:653-658, 1981.

32) Darzynkiewicz, Z., M. Andreeff, F. Traganos and M.R. Melamed. Exp. Cell Res. 115:31-35, 1978.

33) Darzynkiewicz, Z., F. Traganos and M.R. Melamed. Cytometry 3:345-348, 1983.
34) Dolbeare. F., H. Gratzner, M.G. Pallavicini and J.W. Gray. Proc. Natl. Acad. Sci. USA 80:5573-5577, 1983.
35) Darzynkiewicz, Z., F. Traganos, S. Xue, L. Staiano-Coico and M.R. Melamed. Cytometry 1:279-286, 1981.
36) Darzynkiewicz, Z., F. Traganos, S. Xue and M.R. Melamed. Exp. Cell Res. 36:279-293, 1981.
37) Traganos, F., Z. Darzynkiewicz, C. Bueti and M.R. Melamed. Cancer Invest. 2:1-13, 1984.
38) Darzynkiewicz, Z., B. Williamson, E.A. Carswell and L.J. Old. Cancer Res. 44:83-90, 1984.
39) Crissman, H. and J.A. Steinkamp. Cytometry 3:84-90, 1983.
40) Crissman, H.A., Z. Darzynkiewicz and J.A. Steinkamp. Cell Tissue Kinet. 16:521, 1983.
41) Higgins, P., M. Piwnicka, Z. Darzynkiewicz and M.R. Melamed. Am. J. Pathol. 115:31-35, 1984.
42) Johnson, L.V., M.L. Walsh and L.B. Chen. Proc. Natl. Acad. Sci. USA 77:990-994, 1980.
43) Darzynkiewicz, Z., F. Traganos, L. Staiano-Coico. J. Kapuscinski and M.R. Melamed. Cancer Res. 42:799-806, 1982.
44) Darzynkiewicz, Z., F. Traganos, T. Sharpless and M.R. Melamed. Cancer Res. 37:4635-4640, 1977.
45) Darzynkiewicz, Z., F. Traganos, M. Andreeff, T. Sharpless and M.R. Melamed. J. Histochem. Cytochem. 27:478-485, 1979.
46) Kapuscinski, J., Z. Darzynkiewicz and M.R. Melamed. Cytometry 2:201-211, 1982.
47) Darzynkiewicz, Z. In: Effects of Drugs on Cell Nucleus. Bush, H. et al (eds). Academic Press, New York, pp. 270-273, 1979.
48) Roti Roti, J.L. and R. Higashikubo. In: Biomathematics and Cell Kinetics. Rotenberg, M. (ed.) Elsevier, New York, pp. 223-231, 1981.
49) Darzynkiewicz, Z., F. Traganos, T. Sharpless and M.R. Melamed. Proc. Natl. Acad. Sci. USA 73:2881-2886, 1976.
50) Puck, T.T. and J. Steffen. Biophys. J. 3:379-397, 1963.
51) Darzynkiewicz, Z., F. Traganos and M. Kimmel. In: Techniques for Analysis of Cell Proliferation. Gray, G.W., Z. Darzynkiewicz (eds). Humana Press, Clifton, New Jersey (in press).
52) Kimmel, M., Z. Darzynkiewicz, O. Arino and F. Traganos. J. Theor. Biol. (in press).

53) Kurland, J., F. Traganos, Z. Darzynkiewicz and M. Moore. Cell Immunol. 36:318-330, 1978.
54) Brock, W.A., D.E. Swartzendruber and D.J. Grdina. Cancer Res. 42:4999-5003, 1982.
55) Rønning, O.W. and P.O. Seglen.   J. Cell Physiol. 112:19-26, 1982.
56) Pochron, S., M. Rossini, Z. Darzynkiewicz, F. Traganos and R. Baserga. J. Biol. Chem. 255:4411-4414, 1980.
57) Zetterberg, A. and W. Engstrom.   Exp. Cell Res. 144:199-207, 1983.
58) Buroni, D. and C. Ceccarini. Exp. Cell Res. 150:505-508, 1984.
59) Stanners, C.P., M.E. Adams, J.L. Harkins and J.W. Pollard. J. Cell Physiol. 100:127-138, 1979.

ACKNOWLEDGEMENTS

Supported by Grants CA28704 and 23296 from the National Cancer Institute. I thank Dr. Frank Traganos and Miss Robin Nager for their assistance in the preparation of the manuscript.

SECTION IV

NEOPLASTIC TRANSFORMATION

CANCER CELL HETEROGENEITY IN RESISTANCE TO MECHANICAL

TRAUMA IN THE MICROCIRCULATION AS PART OF METASTASIS

HELENA GABOR

Department of Experimental Pathology, Roswell Park

Memorial Institute Buffalo, New York 14563

ABSTRACT Mechanical trauma in the microcirculation
contributes to the inefficiency of the metastatic process
in terms of cancer cells, even though the process progresses and ultimately kills many patients. A filtration
model was developed to assess qualitatively the kinetics of
the damage and survival of cancer cells in passing through
narrow channels, demonstrating that under physiologic
capillary pressures, deformability is a traumatic event
which kills the <u>majority</u> of cells passing through.
Immediate damage to more than 90 percent of cells from the
initial input was internal, reflected in impaired
reproductive integrity (^{3}H-TdR incorporation) and
metabolism (^{14}C-AA incorporation). Analogous loss of
plasma membrane integrity and ultimately cell death were
not apparent until 96h after filtration. A "dormant" state
was observed in the approximately 10 percent surviving
fraction until 214h (L1210 cells) and 240h (EAT cells).
Regular doubling time was resumed thereafter. The amount
of damage and lower survival rates correlated inversely
with the input cell concentration and shear rate and
directly with pore size. Recovery from the trauma of
deformamation was dependent on the spatial association of
the cell and nuclear diameters and the components of the
cytoskeleton and cell periphery. This portion of the work
thus introduces an interesting concept that mechanical
trauma induced in passing through narrow vessels under
physiologic conditions can have a significant effect on the
kinetics of circulating cancer cells. The second part of

this study was designed to determine whether survival from
trauma is random or whether it is a manifestation of a
trauma-resistant subpopulation. However, "wild"
(heterogenous) as well as clonal subpopulations derived
from filtration-trauma survivors were destroyed and
recovered at the same rate as the original "wild" parental
population. Thus, the work supports the concept that
survival from mechanical trauma is a random event and that
resistance is not hereditary.

INTRODUCTION

Based on a histologic analysis of human autopsy data,
Muller (1) considered that circulating cancer cells seldom
survive passage through the first organ encountered, as
metastases were rarely seen in other tissues. Further
studies on human autopsy material have generally supported
the concept that "post-first organ" metastases are less
common than "first organ" metastases (2). The first
studies actually demonstrating cancer cell destruction in
the circulation (3) have shown that after intravenous
inoculation of viable cancer cells, mainly damaged cells
were present in the blood vessels. Furthermore, intra-
venously injected cancer cells are killed at a much higher
rate than the same cells injected by other routes suggesting that the high death rate is related to the mechanical
stress the cells are subjected to in passage through blood
vessels (4). Studies on the kinetics of cancer cell
destruction in the circulation have demonstrated that less
than one percent of the initial intravenous inoculum
actually survives to form metastases, most of the cells
being killed or damaged in transit through the lungs
("first organ") as extrapulmonary metastases are rare
(5,6). The destruction of circulating cancer cells has
generally been attributed to host defense mechanisms of
immunologic and non-immunologic nature (7), the reticulo-
endothelial system (8), and blood coagulation and platelet
aggregation (9). In this communication it is proposed that
nonspecific mechanical trauma is responsible for the
destruction of the majority of cancer cells in the
circulation.

It has been demonstrated that an important mechanical
factor is the trauma due to the repeated deformation in
passage through the microcirculation (10). It appears that
more easily deformable cells are subjected to less trauma
in transcapillary passage; cells which were more easily

deformable *in vitro* were able to pass through the lungs and give rise to postpulmonary metastases after intravenous inoculation. Conversely, resistance to deformation *in vitro* correlated well with a high percentage of cells being killed in one lung passage.

In order to study the phenomena of mechanical trauma in more detail, we must turn to *in vitro* model systems which allow for the variation of selected known factors. Filtration techniques can be used to study whole populations of cells and allow for reproducible alterations of many physical parameters, and microscopic studies of cells in the filters can be made. Thus, qualitative information regarding the mechanisms involved in cell destruction from the trauma of repeated passage through narrow channels may be obtained from filtration studies and may be relevant from a purely *physical* viewpoint to the damage done by passage through the microcirculation. A micropore filtration technique was, therefore, developed for the present study.

It has been suggested that one reason for the numerical disparity in metastasis is that secondary lesions arise predominantly or exclusively from relatively stable, pre-existing minor subpopulations of cancer cells within the primary tumor (11). As the mechanical resistance to circulatory trauma may be a major factor in determining the efficacy of the metastatic process, the second part of this study was designed to determine whether there exists within a population of cancer cells a trauma-resistant subpopulation, and whether this property is heritable.

MATERIALS AND METHODS

Cells and Filtration. EAT and L1210 cells were used from suspension culture and were filtered at concentrations of 0.13 to 1.0 x 10^6/ml. A filtration apparatus was developed, using Nuclepore polycarbonate filters with 5, 8, 10 and 12 μm pores in a Millipore 25 mm ultrafiltration cell. Positive driving pressure of 30 cm H_2O was applied.

Cell Damage and Survival. Damage and survival of cells after passing through filters under standardized conditions were studied by a number of techniques. Cell counts and assessment of plasma membrane integrity were made by dye exclusion. Reproductive capacity was assessd by the ability to divide upon further *in vitro* culture of filtered

cells and by their ability to synthesize DNA by the
incorporation of ^{3}H-thymidine (^{3}H-TdR). Active
cellular metabolism was assessed by the ability of the
cells to synthesize proteins by the incorporation of a
^{14}C-labeled amino acid mixture (^{14}C-AA). Standard
radiolabeling techniques were used.

Isolation of Subpopulations from Filtration-Damage
Survivors (FS). Filtrates were collected after one passage
through 5 µm pores and were either placed in further in
vitro culture immediately ("wild") (12), or adjusted by
end-point dilution to 10 cells/ml and seeded in wells of
microtest plates (clonal). Ten subpopulations of each cell
type were derived from single FS cells (13).

RESULTS

Greatest cell loss occured in the first passage through
filters with 5 µm pores of both cell types at an input concentration of 1x10^6/ml (12). Most pores were obstructed
so that the cells were exposed to the highest shear rates
(12). Therefore, survival was assessed under these conditions and was expressed as percent of the output at t=30
min. (Fig. 1a). Progressive decrease in the percent of
dye- excluding cells was observed in the first 96h after
replacement in culture. Thereafter, no increase in cell
number occurred in the filtrates up to 240h (EAT) and 216h
(L1210). There was a 130±7 and an 84±3 percent increase in
cell number in 24h of unfiltered L1210 and EAT controls,
respectively. A steady percentage of dye-excluding cells
was counted over the time period of > 48h after filtration
(Fig. 1b). The "steady-state" in cell numbers observed
over the period of 96-216h (L1210 cells) and 96-240h (EAT
cells) was not, therefore, due to division of survivors
being balanced by increased cell death, but to a "dormant"
(14) state of the survivors. Exponential growth was
resumed thereafter as a result of division in the 10
percent surviving fraction.

Fig. 2   shows dye exclusion and ^{3}H-TdR and
^{14}C-AA incorporation (a) immediately after one passage
through filters with 5-12 µm pores, expressed as percent of
the input cell number at t=0, and (b) in 24h expressed as
percent of the output at t=30 min. Only L1210 cells are
shown. Internal damage, reflected in the arrest of DNA
synthesis was immediate, followed by perturbation of active

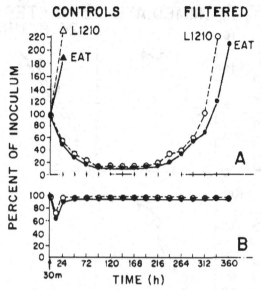

FIGURE 1 Filtration survival of L1210(o) and EAT (•) cells
in suspension culture. A, Number of dye-excluding cells as
percent of filtrate of one passage through 5 μm pores (mean
of 6 experiments, S.E. 3-5%). B, Percent dye-excluding
cells at each interval in 3A.

cellular metabolism before loss of plasma membrane
integrity was apparent. Using these viability assays no
diffrerences were observed in immediate damage as well as
24h survival, between the original "wild" cells and "wild"
populations derived from the 10 percent FS cells.
Thus, subpopulations were cloned from single FS cells and
tested against the original "wild" population. Again, no
differences in destruction and survival rates could be
found (13). The clonal and "wild" populations were then
filtered 5 successive times through filters with 5 μm pores
and the filtrates were then replaced in culture for 72h.
No differences were found in their survival (13).

DISCUSSION

The present study introduces the concept that
mechanical trauma from passage through narrow channels can
have a significant effect on the kinetics of destruction
and survival of cancer cells metastasizing via the blood
stream. It shows that dye exclusion underestimates the

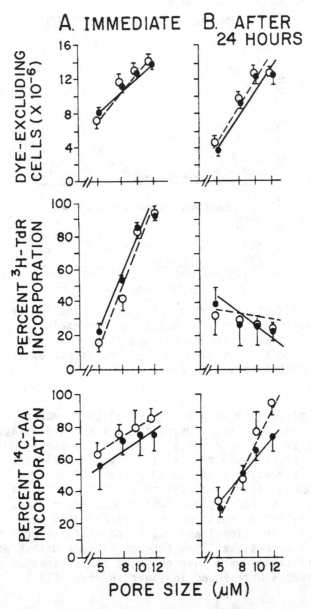

FIGURE 2 Effect of one filter passage on A, immediate cell
damage and B, 24 h survival of previously unfiltered
"wild" L1210 cells (●)   vs. "wild" FS-L1210 cells(o) Pore
size vs. (a) number (mean of 6 experiments ± S.E.) of
dye-excluding cells, (b) ^{3}H-TdR incorporation as percent
(± S.E.) of unfiltered controls, and (c), ^{14}C-AA
incorporation as percent (± S.E.) of controls.

amount of internal damage for several days after trauma has
occurred. The pore size-dependent reduction in ^{14}C-AA
incorporation at both time intervals indicates a
disturbance in active cellular metabolism, such as observed
in cancer cells exposed to cytotoxic agents (14). The
reduction in ^{3}H-TdR incorporation is an indication of
delayed mitosis, i.e., a temporary arrest in the G2 phase
of the cell cycle or a "dormant" state, similar to that
observed in cells exposed to radiation (15). The pore
size-independent reduction in ^{3}H-TdR incorporation in 24h
indicates that damage to the mitotic apparatus occurs prior
to the metabolic damage. Thus, progression of the
different stages of filtration trauma can be demonstrated,
each corresponding to the loss of a specific biologic
function leading either to (a) cell death of the majority
of the cells or (b) a "dormant" state in less than 10
percent of the original input and eventually a restoration
of reproductive capacity.

The question was posed whether the surviving cell
population can resist further filtration trauma; however,
upon refiltration the survivors had no demonstrable inherent resistance to further damage than the original
population. Since numerous studies have considered the
existence of a genetically predetermined, relatively stable
subpopulation of cancer cells from which metastases
predominatly or exclusively arise, one of its major
characteristics, purely from the physical viewpoint, should
be the ability to resist mechanical intravascular
destruction. However, this study presents no evidence
that clonal subpopulations are resistant to further
filtration damage. Thus, based on these findings,
resistance to trauma in passing through narrow channels is
a random event.

This work was partially supported by grant CD-21 from the
American Cancer Society to Dr. L. Weiss. Dr. Gabor's
present address is Children's Hospital, Oakland, CA, 94609.

REFERENCES

1. Muller, J. 1892. Beitrabe zur Kenntnis der
   Metastasenbildung maligner Tumoren. Dissertation Bern.
2. Dukes, C.E. and Busey, H.J.R. 1941. Venous spread in
   rectal cancer. Proc. Royal Soc. Med. 34:571-573.
3. Iwasaki, T   1915. Histological and experimental

observations on the destruction of tumour cells in the
blood vessels. J. Path. Bacteriol. 20:85-105.

4. Hofer, K. G., Prensky, W., and Hughes, W.L. 1969.
Death and metastatic distribution of tumor cells in
mice monitored with ^{125}I-iododeoxyuridine. J. Natl.
Cancer Inst. 43:763-773.

5. Fidler, I.J. 1970. Metastasis: Quantitative analysis
of distribution and fate of tumor emboli labeled with
[125]-I-5-iodo-2'-deoxyuridine. J. Natl. Cancer Inst.
45:773-782.

6. Weiss, L. 1980. Cancer all traffic from the lungs to
the liver: An example of metastatic inefficiency.
Int. J. Cancer 25:385-392.

7. Vaage, J. 1978. In vivo and in vitro lysis of mouse
cancer cells by antimetastatic effectors in normal
plasma. Cancer Immunol. Immunother. 4:257-261.

8. Sadler, T.E. and Alexander, P. 1976. Trapping and
destruction of blood-borne syngeneic leukemia cells in
lung, liver and spleen of normal and leukaemic rats.
Br. J. Cancer: 33:512-520.

9. Colucci, M., Giavazzi, R., Alessandri, G., et al.
1981. Procoagulant activity of sarcoma sublines with
different metastatic potential. Blood 57:733-735.

10. Sato, H., Khato, J., Sato, T., et al. 1977. Deformability and filterability of tumor cells through
"Nuclepore" filter, with reference to viability and
metastatic spread. Gann 20:3-13.

11. Fidler, I.J. and Kripke, M.L. 1977. Metastasis
results from pre-existing variant cells within a
malignant tumor. Science 197:893-895.

12. Gabor, H. and Weiss, L. 1984. The
mechanically-induced trauma suffered by cancer cells in
passing through pores in polycarbonate membranes.
Invasion and Metastasis (submitted).

13. Gabor, H. and Weiss, L. 1984. Survival of L1210 and
Ehrlich ascites cancer cells after mechanical trauma is
a random event. Invasion and Metastasis (submitted).

14. Wheelock, E.F., Weinhold, K.J., and Levich, J. 1981.
The tumor dormant state. Adv. Cancer Res. 34:107-140.

15. Kato, T. and Nemoto, R. 1980. Rapid assay system for
cytotoxicity tests using ^{14}C-leucine incorporation
into tumor cells. Tohoku J. Exp. Med. 1131:261-270.

16. Sapareto, S.A., Raaphorst, G.P., and Dewey, W.C.
1979. Cell killing and the sequencing of hyperthermia
and radiation. Int. J. Radiat. Oncol. Biol. Phys.
5:343-347.

EXPRESSION OF AN "ACTIVATOR" RNA HYBRIDIZABLE TO THE SV40 PROMOTER CORRELATES WITH THE TRANSFORMED PHENOTYPE OF MOUSE AND HUMAN CELLS.

Margarida Krause, Jolanta Kurz and Uik Sohn

Department of Biology, University of New Brunswick, Fredericton, New Brunswick, E3B 6E1 Canada.

INTRODUCTION

For the past seven years our laboratory has been engaged in the study of a small molecular weight nuclear RNA-7S-K, whose properties support the concept that it is a gene regulatory molecule (8-10, 12-14). This RNA was found to stimulate transcription initiation by RNA polymerase II (13) and to act in a tissue- and species-specific manner (10). On the basis of these results we hypothesized that different subpopulations of 7S-K RNA might exist in different tissues and that in each case, they would be capable of recognizing promoters of different gene families, facilitating the formation of the initiation complex for transcription of tissue-specific genes. This hypothesis was subsequently tested in simian virus 40 (SV40)-transformed mouse 3T3 cells in which the SV40 early gene for T-antigen was found to be actively transcribed and necessary to maintain the expression of the transformed phenotype (15). These cells were selected because the SV40 genome has been completely sequenced and its promoter has been better characterized than any other eukaryotic promoter (7). These recent experiments have shown that 7S-K RNA from SV40-transformed mouse cells does indeed hybridize to the SV40 early promoter, with an extensive homology with two of the 21 base-pair repeats(14). Two of the three repeats have been identified as minimum essential elements for early viral promoter function (3,6).

The present study was designed to investigate if the degree of homology between 7S-K RNA and the SV40 promoter

correlates with the expression of the transformed phenotype. To do this we used a variety of human and mouse cell
lines ranging from embryonic primary cells, which are still
diploid and subject to aging, to immortalized established
lines, which are still contact inhibited and nontumorigenic,
to fully transformed tumorigenic lines obtained either by
SV40-transformation or from a malignant tumor. However,
when comparing normal and transformed cell lines, one must
bear in mind that, even when they originate from the same
parent strain, they may have subsequently diverged in ways
unrelated to the process of transformation per se. To
circumvent this problem, we also used a temperature
sensitive mutant of SV40 (SV40 tsA), which is defective in
the large T antigen gene, to transform 3T3 mouse cells.
Cells thus transformed express the transformed phenotype
at 33°C (permissive temperature) and revert to normal at
39°C (nonpermissive temperature) (1). This allows for the
manipulation of phenotypic expression of the transformed
state by a simple shift in culture temperature.

MATERIALS AND METHODS

All cell lines were cultivated in Dulbecco's modified
Eagle's medium supplemented with 10% fetal or newborn calf
serum. Small RNAs were isolated as described (13) and
fractionated in 6-15% polyacrylamide gradient gels or
3-12% polyacrylamide-7M urea gels. Gels were stained with
0.5 mg/liter ethidium bromide and photographed under UV.
For Northern blot hybridization the RNA in the gels was
electrotransferred to DBM paper (2). The probes used
(mSV01 + mSV02) are M-13 phages containing the 311 bases
SV40 EcoRII-G fragment in either orientation (a gift from
Dr. M. dePamphilis). The replicative form of the phage was
labeled with [^{32}P] dCTP by nick translation (11). Hybridizations were carried out in 50% formamide 0.75 M NaCl/75 mM
Na Citrate (5 x SSC) 25 mM Na2PO4 (pH 6.5)/0.1% each bovine
serum albumin, ficoll and polyvinylpyrrolidone containing
250 µg/ml denatured salmon sperm DNA for 48 hr at 50°C
followed by brief washes, once in 2xSSC/0.5% SDS, once in
2xSSC/0.1% SDS at r.t. and a 2-hr wash in 0.1xSSC/0.5% SDS
at 50°C respectively.

RESULTS AND DISCUSSION

Isolation and fractionation of small RNAs from nuclear
and cytoplasmic fractions of cultured mouse and human cells

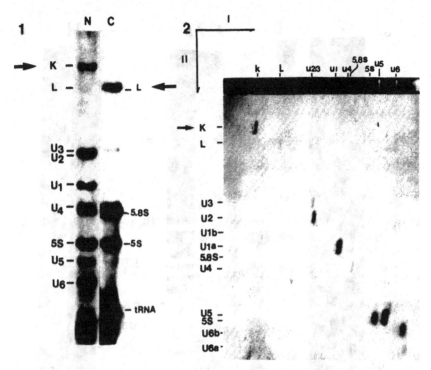

Figure 1. Nuclear (N) and cytoplasmic (C) small RNAs from mouse SV3T3
cells. Cells were labelled in vivo with ^{32}P for 15 hr in phosphate-free
medium and the small RNAs isolated from nucleus and cytoplasmic fractions
(13) fractionated in 5-15% acrylamide gels and autoradiographed.
Figure 2. Two-dimensional separation of small nuclear RNAs from mouse
SV3T3. First dimension 5-15% polyacrylamide, stained with ethidium bromide;
second dimension 3-12% polyacrylamide 7-M urea, silver stained.

has revealed that the species responsible for activating
transcription - 7S-K occurs exclusively in the nuclear
fraction (see Fig. 1). Another 7S RNA-L, which has recently
been found to be an essential component of the signal
recognition particle (16) is a cytoplasmic species which
can occasionally appear as a contaminant in nuclear
preparations. 7S-K nuclear RNA is a 330 nucleotide long RNA
which migrates as a single species in both one-dimension
nondenaturing gels (Fig. 1) and two-dimension gels in which
7M urea is used in the 2nd dimension (Fig. 2). Yet its
properties as a transcription factor vary from tissue to
tissue and are abolished in a different species (10)
indicating that sequence heterogeneity does exist.
        Another variable is the yield of 7S-K RNA from

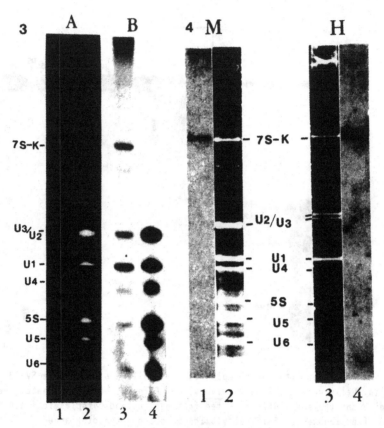

Figure 3. Effect of actinomycin D (AMD) on recovery of 7S-K RNA.
Mouse SV3T3 cells were labelled with ^{32}P as in figure.1 and treated with
0.5 µg/ml AMD during the last 3 hrs (lanes 1 & 3) or untreated (lanes 2
and 4). A-ethidium bromide stained gel; B-autoradiograph.
Figure 4. Exclusive hybridization of 7S-K RNA from mouse SV3T3 cells (M)
and human HeLa cells (H) with the SV40 "Ori" probe. RNAs were
electrotransferred from acrylamide gels to DBM and hybridized with nick
translated probe as described in methods. Lanes 2 and 3 - stained gel
tracks; 1 and 4 - autoradiographs of DBM paper after hybridization.

different cell lines or even from the same cell line at
different stages of growth. We found that, on a per cell
basis, roughly twice as much 7S-K RNA is recovered from
transformed than from nontransformed exponentially growing
cells and four times as much than from contact-inhibited cells.
Furthermore in order to achieve a consistent optimum yield
of 7S-K, we found that the cells must be treated with 0.5
µg/ml actinomycin D (AMD), 1-3 hr before harvest (see Fig.3)

This is because 7S-K tends to leak out of the nucleus
during cell lysis and it is readily degraded. Treatment with
high doses of AMD freezes it in the chromatin, presumably
as part of the transcription initiation complex, when
elongation is inhibited. Note that a substantially larger
quantity of 7S-K is recovered in the actinomycin D-treated
sample (tracks 1 & 3) even though the total amount of
sample loaded is much less than in the control sample
(tracks 2 & 4). The autoradiographs of the RNAs after
in vivo labelling with [^{32}P] orthophorphate show excellent
labelling of 7S-K in spite of the AMD treatment. This is
not surprising not only because these small RNAs have long
half lives but also because they arise from Class III genes
which are highly resistant to the drug.

Our previous report that 7S-K RNA from SV40-trans-
formed mouse cells has an extensive homology (45 bases)
with the SV40 promoter (14) raised the question of whether
or not this is an exclusive property of SV40-transformed
cells and or is somehow related to transformation. At the
time, the hybridization signals obtained were rather weak,
due in part to the low and variable yields of 7S-K and to
the low yield of the EcoRII fragment "G", which contains
the early promoter region and which must be recovered from
the agarose gel after EcoRII digestion of total SV40 DNA
before nick translation and hybridization with the small
RNAs. Because of this no signals were detected with the RNAs
of untransformed mouse cells. Once we optimized the yield
of 7S-K by AMD treatment and obtained the M13 clones with
the EcoRII "G" insert (Ori region), the protocols were
greatly simplified and the strength of the hybridization
signals was substantially increased. Figure 4 shows that
both mouse and human cells, the first SV40-transformed, the
second obtained from a human tumor, have homology between
7S-K RNAs and the SV40 promoter. No other nuclear or
cytoplasmic small RNAs was found to hybridize to the M13
cloned fragment. We then tested a number of cell lines and
found that untransformed mouse and human cells also show
hybridization with SV40, albeit at a lower level. Because
the relative amount of K RNA varies from one cell line to
the next, it was necessary to quantitate, in each case, the
total amount used for hybridization and to relate this value
to the hybridization signal. This was done by densitometer
analysis of both ethidium-bromide stained gels and the
respective autoradiographs obtained after transfer to DMB
and hybridization with the probe. Table I illustrates the
results. It is apparent that transformed tumorigenic cells

Table I

Correlation between the degree of hybridization between 7S-K RNA and SV40
"Ori" region and cancerous features of mouse and human cell lines.

Cell line	transformation	Tumorigenic	hybridization ratio[1]
mouse 3T3	established line	No	1.0
mouse L-929	established line	No	1.5
mouse SV3T3	SV40	Yes	2.5
mouse A255 (non-permissive temp.)	SV40 (tsA)	No	1.0
mouse A255 (permissive temp.)	SV40 (tsA)	Yes	2.3
human WI38 (diploid, embryonic lung)	primary line	No	0.7
human SVWI38	SV40	Yes	1.4
human heLa	established line	Yes	4.4

[1]the hybridization ratio was obtained by densitometer scannings of both
ethidium bromide stained gel tracks of 7S-K and the respective
hybridization signal obtained in the X-ray. Those were computer analysed
and the ratio of hybridized signal over stained gel signal were obtained
and normalized to 1.0 for mouse 3T3 cells. Standard deviations between
experiments did not exceed 5%.

have twice as much RNA hybridizable to the SV40 early
promoter than their respective parent strains. That this
difference is related to transformation rather than to
circunstantial divergence between cell lines can be deduct-
ed from the finding that the A255 cell line, which is
temperature-sensitive to expression of transformation,
expresses twice the amount of 7S-K RNA, hybridizable to the
SV40 promoter, at the permissive temperature. Moreover,
tumorigenic characteristics of the cells, rather than the
presence of SV40 genes, appears to be the key factor in
this relationship with SV40-hybridizable small RNA, since
highly tumorigenic HeLa cells (derived from a human
cervical carcinoma) show the strongest signal.
      Previous results from our laboratory have shown that
7S-K is transcribed from a midrepetitive family of host
sequences (14) Screening of a recombinant library of
human DNA with the same SV40 "Ori" probe have yielded
large numbers of clones hybridizable to the probe. These,
in spite of being partially homologous to one another, were
not found to be part of a highly conserved family of

sequences and were not members of the previously described
"Alu" family of repeats (4). The hybridized region of one
of these clones (SVCR7) showed the same 6 sequence motifs
(CCGCCC) that were found to be essential elements in the
21 base pair repeats of the SV40 promoter (6). More recent
work on promoter-specific factors have revealed that one
such factor isolated from HeLa cells (SpI), is a promoter-
specific factor which activates a class of promoters that
includes the SV40 early promoter but not several others
that have been tested (5). Analysis of promoter deletion
mutants and DNAse footprinting assay revealed that SpI
binds to the 21 bp regions of SV40 (5). The similarity
between these findings and ours suggests that 7S-K RNA is
part of the SpI factor,helping in selective activation of
a class of promoters which may participate in transformation. By base-pairing to the late strand, in the case of
the SV40 genes, (14) it may actually determine the
direction of transcription since the template (early strand)
is then free to be copied into early messenger RNA, which
is translated into T-antigen, a protein needed to maintain
SV40 transformation (15).

Supported by the Natural Sciences and Engineering Research Council of
Canada, the National Cancer Institute of Canada and the Terry Fox Grant
Fund of New Brunswick.

REFERENCES

1. ALWINE, J.C., S.I. REED and G.R. STARK, 1977. Characterization of the
   autoregulation of simian virus 40 gene A. J. Virol. 24, 22-29.
2. ALWINE, J.C., D. KEMP and G. STARK. 1977. Method for detection of
   specific RNAs in agarose gels by transfer to diazobenzyloxymethyl
   paper and hybridization with DNA probes. Proc. Natl. Acad. Sci. 74,
   5350 5354.
3. BYRNE, B.Y., M. DAVIS, J. YAMAGUCHI, D.J. BERGSMA and K.N.
   SUBRAMANIAN. 1983. Definition of the simian virus 40 early promoter
   region and demonstration of a host range bias in the enhancement
   effect of the simian virus 40 72-base pair repeat. Proc. Natl. Acad.
   Sci. U.S.A. 80, 721-725.
4. CONRAD, S.E. and M.R. BOTCHAN. 1982. Isolation and characterization of
   human DNA fragments with nucleotide sequence homologies with the
   simian virus 40 regulatory region. Mol. Cell. Biol. 2, 949-965.
5. DYNAN, W.S. and R. TJIAN. 1983. The promoter-specific transcription
   factor SpI binds to upstream sequences in the SV40 promoter. Cell 35
   79-87.

6. EVERETT, R.D., BATY, D. and CHAMBON, P. 1983. The repeated GC-rich
   motifs upstream from the TATA box are important elements of the SV40
   early promoter. Nucl. Acids Res. 11, 2447-2646.

7. JONGSTRA, J., T.J. MATHIS and P. CHAMBON. 1984. Induction of altered
   chromatin structures by simian virus 40 enhancer and promoter
   elements. Nature 307, 708-717.

8. KRAUSE, M.O. AND M.J. RINGUETTE. 1977. Low molecular weight nuclear
   RNA from SV40-transformed WI38 cells; effect on transcription of
   WI38 chromatin in vitro. Biochem. Biophys. Res. Comm. 76, 796-803.

9. KRAUSE, M.O. and M.J. RINGUETTE. 1982. Stimulation of transcription
   in isolated mammalian nuclei by specific small nuclear RNAs. In
   genetic Expression in the cell cycle. Eds. Padilla, G.M. & McCarty
   Sr. K.S. (Academic Press New York) pp. 151-179.

10. LIU, W.C, R. GODBOUT, E. JAY, K.K. YU AND M.O. KRAUSE. 1981
    Tissue and Species specific effects of small molecular weight
    nuclear RNAs on transcription in isolated mammalian nuclei. Can. J.
    Biochem. 59, 343-352.

11. RIGBY, P.W.J., M. DIECKMAN, C. RHODES and P. BERG. 1977. Labelling
    deoxyribonucleic acid to high specific activity in vitro by nick
    translation with DNA polymerase I. J. Mol. Biol. 113, 237-251.

12. RINGUETTE, M.J., W.C. LIU, E. JAY, K. K.Y YU and M.O. KRAUSE. 1980.
    Stimulation of transcription of chromatin by specific small nuclear
    RNAs. Gene 8. 211-224.

13. RINGUETTE, M.J., K. GORDON, J. SZYSZKO and M.O. KRAUSE. 1982.
    Specific small nuclear RNAs from SV40-transformed cells stimulate
    transcription initiation in nontransformed isolated nuclei. Can. J.
    Biochem., 60, 252-262.

14. SOHN, U., J. SZYSZKO, D. COOMBS and M.O. KRAUSE. 1983. 7S-K nuclear
    RNA from simian virus 40-transformed cells has sequence homology to
    the viral early promoter. Proc. Natl. Acad. Sci (USA) 80, 7090-7094.

15. TEGTMEYER, P. 1975. Function of simian virus 40 gene A in transform-
    ing infection. J. Virol. 15, 613-618.

16. WALTER, P. and G. BLOBEL. 1983. Disassembly and reconstitution of
    signal recognition particle. Cell 34, 525-535.

GASTROINTESTINAL CELLS: GROWTH FACTORS, TRANSFORMATION, AND MALIGNANCY

M.P. Moyer, P. Dixon, D. Escobar, J.B. Aust

The University of Texas Health Science Center at San Antonio

7703 Floyd Curl Drive, San Antonio, TX 78284

INTRODUCTION

At the cellular level, growth regulation and differentiation of normal and malignant gastrointestinal (GI) epithelium are poorly understood. This is due to many factors, including previous difficulties in culturing GI cells in vitro for a useful time period (1-4). That problem has been rectified by our development of alternative methods for culturing normal human and animal GI cells (5,6) and GI tumor cells (7). In addition, our recent successful transformation of human colon cells (8) with the chemical carcinogen, azoxymethane, and the oncogenic virus simian virus (SV40) provides a baseline for expanding studies of in vitro transformation to include alternate sources of GI cells. In this paper, we relate some recent studies of in vitro transformation, growth, and characterization utilizing human and rodent GI cells.

MATERIALS AND METHODS

Normal and malignant human and rodent (rat or mouse) GI cells were initiated in vitro as previously described (5-8). Normal cells were harvested from the mucosal epithelium of the stomach, small intestine, and colon. Several base culture media (obtained from M.A. Bioproducts or GIBCO) were included in these studies:

297

minimal essential medium (MEM), Dulbecco's modified MEM
(D-MEM), suspension culture MEM (S-MEM), L15, F12, McCoy's
5A, and CMRL 1066.   Media additives tested were fetal
bovine serum   (FBS; 2-10% [v/v]); insulin   (Sigma;
1-5 µg/ml), transferrin (Sigma; 1-5 µg/ml), pentagastrin
(25 ng - 25 µg/ml; Peptavlon [Eli Lilly] and epidermal
growth factor (EGF; 1-10 ng/ml; Collaborative Research
Inc.). Conditioned medium (25% v/v) from the same culture
was used for all subsequent subcultures.

Morphology and growth were analyzed by direct
observation   and   standard   growth   assay   methods.
Transepithelial transport was indicated by dome formation
(9).   Immunofluorescence   assays   with   commercially
available antibodies were done to analyze the presence of
keratin, carcinoembryonic antigen (CEA) and fibronectin.
Intestinal calcium binding protein (CaBP) and lactase were
assayed by indirect immunofluorescence, using monospecific
primary antibodies (kindly supplied by Drs. W.P. Gleason
and R. Montgomery, respectively). Mucins were assayed by
standard   Periodic   acid-Schiff   (PAS)   staining   and
radioimmunoassay for specific mucosal antigens (kindly
performed by Dr. David Gold). Normal human or rodent GI
cells from the various organ sites (stomach, small
intestine, and colon) were transformed in vitro with
azoxymethane or SV40 virions by methods detailed by Moyer
and Aust (8).

RESULTS

Many factors were important for successful initiation
of primary cultures, which could then be subcultured for
several generations (Table 1). The cultured GI epithelial
cells from humans and rodents displayed common growth
(Table 2) and morphological features (Table 3), and
characteristics specific to GI epithelial cells (Table 4).
Of interest, was that GI cells from the same organ site,
but different species, were similar with regard to
morphology, cell subpopulations, and other phenotypic
features. Growth factors developed for cultures of normal
human GI cells proved suitable for rodent GI cells (Table
5). However, the rodent cells were less fastidious, often
able to grow in medium supplemented with serum only or
lower concentrations of growth factors (data not shown).

Table 1. FACTORS IMPORTANT FOR GI CELL CULTURE OF EPITHELIAL CELLS

Propensity of cells to grow in suspension

Need for mechanical harvesting; standard dissociating chemicals are cytotoxic and often select for fibroblasts

Many rinses of processed tissue must be done

Small size of cells relative to cultured fibroblasts and most cell lines

Growth factors from yeast and brain (or pituitary) extracts optimize growth

"Split ratios" could not exceed 1:4 and conditioned medium required upon subculture

The cells are "delicate" in vitro, requiring time to adapt and careful observation for proper maintenance

---

Table 2. GROWTH CHARACTERISTICS OF CULTURED GI CELLS

Preference to grow in suspension as cell clusters

Differentiation from round or cuboidal dividing cells in cluster to columnar epithelial cells

Growth rate: population doubling time of 72 to 96 hours or more

Variable proportions of dividing cells

Subculture frequency variable: once every 1 to 4 weeks

No growth of single cells in soft agar

Table 3. MORPHOLOGY OF GI EPITHELIAL CELL CULTURES

Clusters of dividing and differentiated cells;
often gland-like structures

Few or no fibroblasts in cultures

Small; $\leq$ 10-12

Columnar and goblet cells with evident
baso-lateral differences

Junctional plasma membrane complexes
typical of epithelial cells

Table 4.   DIFFERENTIATION/EPITHELIAL CELL CHARACTERISTICS DISPLAYED BY HUMAN OR RODENT GI CELLS

Keratin

Carcinoembryonic antigen (CEA)

Little or no fibronectin

Responsiveness to GI hormones

Mucins: Synthesis of specific mucosal antigens; PAS
reactivity
Transepithelial Transport (e.g., dome formation)

Intestinal calcium binding protein

GI-specific enzymes (e.g., lactase)

Growth requirements varied for the large numbers of human GI (primarily colon and stomach) tumors initiated in our facilities. Some grew well in "standard" culture media such as MEM, D-MEM, or L15 supplemented with 10% FBS, whereas others grew best in the media utilized for normal cells (Table 5). Cell lines (initiated in our laboratory or obtained from culture collections) always grew optimally in the original medium used for selection. This included replicates initiated as primary cultures in several types of media. Several observations were consistently noted: (1) the cells grew best in the medium used for primary culture initiation; (2) tumor cell cultures initiated in medium of "high" calcium concentrations (i.e., 1 to 2 mM) and supplemented with 10% FBS consisted of larger cells; these were often contaminated with fibroblasts that overgrew the tumor cells (particularly when standard trypsin-EDTA dissociation was used for subculture); and (3) although some tumor cells preferred the enriched medium used for culturing normal cells, most cultures could be grown in media with fewer growth factor supplements.

Cell cultures and lines of various GI and other human solid tumors were initiated at a high success rate (Table 6), approximately 88% for GI tumors, using methods described elsewhere (7). Lines could be selected for growth in suspension or monolayers, depending on culture media and conditions. Similar observations have been made with rodent solid tumors.

As previously reported for human colon epithelial cells (8), SV40 and azoxymethane transformed a variety of GI cell cultures from the stomach, small intestine (duodenum, jejunum and ileum) and colon from humans and rodents. General characteristics displayed by the transformants are shown in Table 7. Of particular note was that some features (e.g., growth in soft agar) are also reported as "transformation-associated" charac-teristics of fibroblasts, whereas others (e.g., enhanced substrate adherance) are opposite to observations of in vitro transformed fibroblasts.

Table 5. COMBINATION OF FACTORS STIMULATING GROWTH OF
GASTROINTESTINAL CELLS

Growth Factor	Concentration
FBS	2% (v/v)
L Broth*	2% (v/v)
Pituitary extract	1% (v/v)
Insulin	5 μg/ml
Transferrin	5 μg/ml
Selenium	5 ng/ml
Hydrocortisone or	10 μg/ml
Dexamethasone	40 ng/ml
Pentagastrin	25 μg/ml
EGF	10 ng/ml
Conditioned Medium	25% (v/v)

*L Broth: provides source of yeast extract; composition:
NaCl (5 g/l); bactotryptone (10 g/l); yeast extract
(5 g/ml)

Table 6. CULTURES OF COLON CANCERS AND OTHER
SOLID HUMAN TUMORS INITIATED IN OUR LABORATORIES

Site or Type	No. Cultured/ No. Attempted	Percent Cultured
Colorectal	34/39	87%
Gastric	8/9	89%
Basal Cell	7/7	100%
Pancreatic	5/6	83%
Insulinoma	1/1	100%
Bladder	2/2	100%
Mammary	2/3	67%
Ovarian	1/1	100%

Table 7. EPITHELIAL CELL CHARACTERISTICS CHANGED UPON IN VITRO TRANSFORMATION BY SV40 OR AZOXYMETHANE

More cells adhere to culture substrate

Decreased requirement for growth factors
in yeast or brain extracts or conditioned medium

Increased cell size

Decreased sensitivity to dissociating chemicals

Increased culture longevity

Acquisition by some cells of ability to grow in soft agar

Cell surface change: binding of peanut agglutinin lectin

SV40 transformants: positive for SV40 T antigen

DISCUSSION

Successful culture of human and rodent cells confirm and expand previous studies (5). Display of morphological, epithelial, structural, functional, and intestinal differentiation characteristics indicated that the methods for culture initiation and propagation were suitable, although efforts continue to further improve the culture environment. Interestingly, the rat cells were less fastidious than the human cells in their growth requirements, particularly after multiple subcultures. This concurs with studies of fibroblastic cells that show rodent cells are more easily propagated, have greater longevity, and are more readily established as cell lines, than human cells. However, cell lines have not yet been selected from any of the normal rat or human intestinal cells.

Likewise, no cells lines have been selected from any of the in vitro transformed cultures, although they were maintained for multiple subcultures beyond the controls. In contrast, cell lines have been selected in our

laboratory and others from solid gastrointestinal tumors.
Improved "culturability" of the transformed and/or tumor
cells is probably explained by auto-stimulation from their
own growth factors, as reported in other systems (10).

The other phenotypic changes observed in the
transformed cells indicate an intermediate phenotype
between normal and tumor cells. These changes are only a
subset of GI cell transformation-induced alterations.
Many more remain to be defined. Ongoing efforts to better
characterize the transformants and to compare normal and
tumor cells include studies of lectin binding, monoclonal
antibodies, and a panel of biochemical/cytological
features. Also in progress are transfection studies with
oncogenic virus DNAs maintained as recombinants with
dominant selectable markers. In that regard, recent
observations indicate that transformation of human and rat
GI cells has been achieved (Moyer, unpublished
observations) with pSV3gpt, an SV40 early gene recombinant
with a gpt selectable marker (11).

Future work in developing these in vitro GI models
will provide a valuable foundation in understanding GI
malignancies, and will better define factors regulating GI
physiology, toxicology and differentiation.

REFERENCES

1. Franks LM. 1976. Cell and organ culture techniques
   applied to the study of carcinoma of the colon and
   rectum. Pathol Eur 11:167-177.

2. Franks LM, Wilson PD. 1977. Origin and
   ultrastructure of cells in vitro. Int Rev Cytol
   48:55-139.

3. Quaroni A, May RJ. 1980. Establishment and
   characterization of intestinal epithelial cell
   cultures. Methods Cell Biol 21:430-427.

4. Smith HS. 1979. In vitro properties of epithelial
   cell lines established from human carcinomas and
   nonmalignant tissue. J Natl Cancer Inst 62:225-230.

5. Moyer, MP. 1983. Culture of gastrointestinal
   epithelial cells. Proc Soc Exp Biol Med 174:12-15.

6. Moyer MP, Page, Moyer RC. 1984. In vitro culture of
    gastrointestinal epithelial cells and tissues. In:
    Webber M, Sekeley L, eds. In Vitro Models of Human
    Cancer. Boca Raton, CRC Press Reviews.

7. Moyer MP. 1984. A rapid, reproducible method for
    processing human solid tumors for in vitro culture.
    J Tissue Culture Methods (in press).

8. Moyer MP, Aust JB. 1984. Human colon cells:
    Culture and in vitro transformation. Science (in
    press).

9. Lever JE. 1982. Cell culture models to study
    epithelial transport. In: Martonosi (ed).
    Membranes and Transport. Plenum Publishing Corp,
    pp. 231-236.

10. De Larco JE, Todaro GJ. 1979. Sarcoma growth
    factor: Specific binding to and elution from
    membrane receptors for epidermal growth factor. Cold
    Spring Harbor Symp Quant Biol 44:643-649.

11. Mulligan RC, Berg P. 1981. Selection for animal
    cells that express the Escherichia coli gene coding
    for xanthine-guanine phosphoribosyltransferase. Proc
    Natl Acad Sci 78:2072-2076.

ACKNOWLEDGMENTS

Research partially supported by the Morrison Trust,
and Institutional Research Grants from the NIH (RRO5654)
and the American Cancer Society.

A NEW TECHNIQUE FOR DIAGNOSING CANCER BY

INSPECTING BLOOD SERUM

A. Kovács, A. Vértesy, L. Szalai,

S. Adámi, L.Urbancsek, Z. Simon,

G. Németh, J.M. Takács

Cancer Research Group, c/o A.Kovács

Buza utca 2., H-2097 Pilisborosjenő
Hungary

Abstract

A new method has been established for identifying
cancer by inspecting the blood serum. This
technique is based on the perception that there
is a marked difference in the composition of
protein in the normal human blood serum and the
deseased one. This novel technique is extremely
beneficial to quick cancer checking of large
groups of patients.

Examinations which are focused on the human
blood serum play an ever increasing role in
medical diagnostics. Very much information of
fundamental importance can be gathered by analys-
ing the results of blood tests.

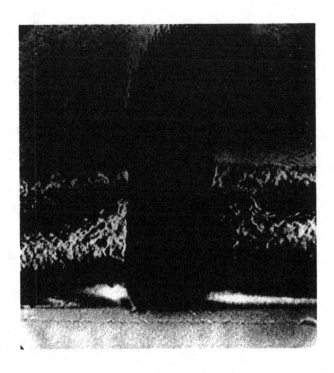

Fig. 1

The characteristic picture of dried
blood-serum with threads (taken by
scanning electron microscope, N=1500)

We have not come across international publications
dealing with the morphology of the dried blood
serum so far. During our investigation we had
been studying in morphological and topological
respect the dried blood serum taken from both
normal and ill patients suffering from tumorous
illness.

Suitably prepared blood serum dropped on a glass
plate is left to dry at room temperature. After

the drying process is completed a characteristic
picture develops which can be observed by an
optical microscope. The dried blood serum becomes
crackled and in the splits threads consisting of
nodes can be seen. Besides, on the surface of the
smear various kinds of salt condensations develop.
This is shown in Fig.1 and 2.

During our work we had been studying the drying
process, the nature and the synthesis of the
treads. It has been found that the threads are
formed at first and they influence the splitting

Fig. 2

Detail of a thread
(SEM, N=6000)

of the blood serum. Before the drying process was
completed a piece of blood-serum was removed from
the middle of the drop. At this phase splits can
not be observed yet on the surface but the threads
and their nodes have been developed at the bottom.
The next task was to determine the synthesis of
the threads. It seemed to be obvious that the
threads originate from plasma protein. Na_2SO_4
solution added to a fresh drop of blood serum
prevents the formation of the thread structure.
Namely, the Na_2SO_4 solution precipitates the
proteins in the blood serum which in turn, can not
become arranged. After having had found this indirect
proof the composition of the threads was investigated by means of analytical chemistry. Proteins in
the threads interact and this interaction is
responsible for the structure of the net.

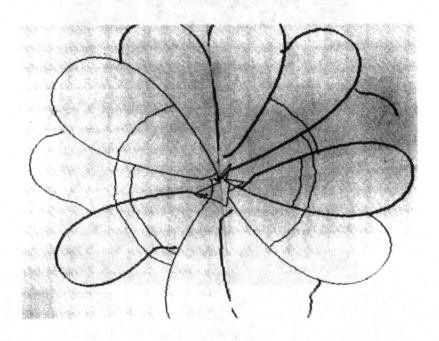

Fig. 3

The drying process of the normal and cancerous
blood serum was modelled by simulation programs
which ran on a PDP-11 computer system. The programs
were written in BASIC.

In case of normal samples it was assumed that the
threads proceed straight from their starting points
scattered at random. This process is unchanged as
long as two threads get near "enough" to each
other. Then the mutual attractive force between
them diverts the one which grows later on.

As for tourous samples it was supposed that there
were very strong interactions in the medium which
generate more starting points per area than in
normal cases, on the one hand and modify the progression of the threads at random on the other.

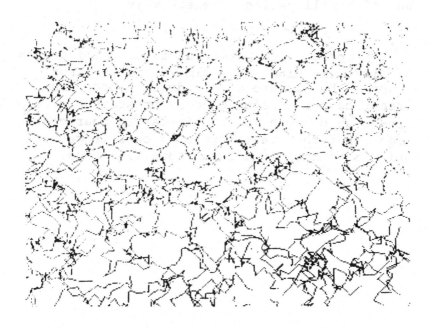

Fig. 4

Figures 3 and 4 show the pictures of the plot
generated by the simulator described above.

Pattern recognition and digital image processing
programs are of vital importance in performing
quick cancer checking of large groups of patients.

More than two thousand blood serum drops taken
from different groups of patients have been investigated so far. Among these, many patients
suffered from tumorous illness. The thorough
visual analysis of the surface of dried blood-
serum drops has shown that there is a marked
difference between that of the normal (or human
beings suffering from other illness than tumorous)
and the cancerous patient. Figures 5 and 6 show
the image of the dried blood serum of a healthy
and of an ill patient respectively.

It is important to note that tumorous illnesses
of different kinds show similar images. For
example the picture of the dried blood-serum of
breast cancer, rectum cancer, prostate cancer and
Non-Hodgkin lymphoma are similar in character. It
means that the plasma proteins carry something
which is common in all tumorous illnesses.

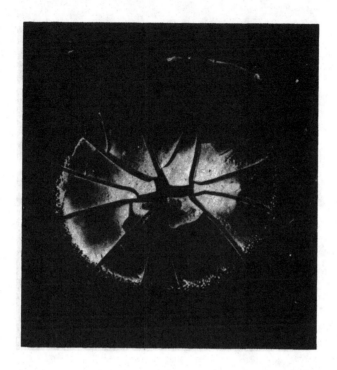

Fig. 5

The picture of the dried blood serum
of a patient who does not suffer from
tumorous illness. (SEM, N=20)

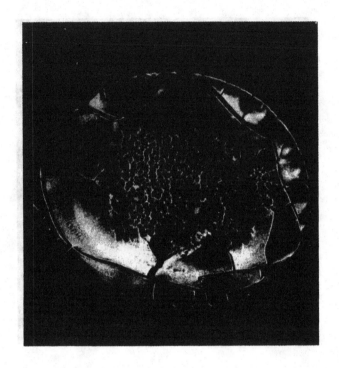

Fig. 6

The image of the dried blood serum of
a patient who suffers from tumorous
illness. (SEM, N=20)

CYTOKINETICS OF HETEROPLOID TUMOR SUBPOPULATIONS BY COMBINED AUTORADIOGRAPHIC IMAGING AND FEULGEN DENSITOMETRY.

Robert J. Sklarew, Ph.D.

New York University Research Service
New York University School of Medicine
Goldwater Memorial Hospital
Roosevelt Island, New York, NY 10044

Traditional autoradiographic methods for analysis of the cell cycle kinetics of tumors have been limited to estimation of mean phase transit-times and their dispersion in the population as a whole. This obscures the proliferative heterogeneity which may exist among subpopulations. In human tumors where polyploidy and aneuploid variants frequently co-exist, such measurements do not relate to the cytokinetic behavior of the distinct ploidy lineages. An important objective has therefore been to resolve their independent proliferative contributions. This has posed some formidable problems: In populations of uniform ploidy the cycle phase distribution may be resolved from statistical treatment of DNA distributions obtained by flow cytofluorometry or scanning densitometry (1). However, in mixed polyploid/aneuploid systems such a phase deconvolution of subpopulations is not amenable to precise methods due to extensive overlap of their DNA content ranges in the various cycle phases. It is the purpose to outline an approach for resolving the phase distribution and compartmental turnover kinetics of subpopulations based upon automated image-analysis of ^{3}HTdR labeling and Feulgen densitometry in autoradiographs. The technical aspects of the methodology have been reported (4-6). The scheme is an extension of principles described for deconvolution of the S-phase compartment in heteroploid cell populations (7,8)

MATERIALS AND METHODS

315

Labeling of Cell Cultures

A heteroploid subline of MCF-7 human breast cancer cells (3) was cultured at 37°C in Eagles MEM Medium containing 10% fetal calf serum L-glutamine (292mg/l), neomycin (50mcg/ml), and insulin (10 mcg/ml). Cultures were split 2:1 every 10-14 days and seeded at 4 x 10^5 cells/ml in Falcon T-250 flasks in 25 ml of media. Experiments were initiated in 7 day log phase cultures. Feulgen-stained autoradiographs were prepared after Carnoy-fixation (4).

I. ^{3}H- and ^{14}C-Thymidine double-labeling. Replicate cultures were pulsed for 20 min with ^{3}H-thymidine (0.1 µCi/ml, Sp. Act 6.7 Ci/mM), grown in Colcemid (0.05 mcg/ml) for up to 18h and then pulsed with ^{14}C- thymidine (0.1µCi/ml, Sp Act 53 Ci/mM) 20 min prior to fixation.

II. ^{3}H-Thymidine continuous-labeling. Replicate cultures were exposed continuously to ^{3}H-Thymidine (0.01 µCi/ml., Sp. Act. 6.7 Ci/mM) for up to 18h in the presence of Colcemid (0.05 mcg/ml). Prior to termination they were pulsed for 20 min. with ^{14}C-thymidine (0.1µCi/ml, Sp. Act. 53Ci/mM).

III. ^{3}H-Thymidine pulse-labeling. Replicate cultures were pulsed with 3-Thymidine (0.1 µCi/ml., Sp. Act. 6.7 Ci/mM) for 20 min and fixed. Diploid rat kidney cultures were pulsed and processed in parallel to provide 2C and 4C DNA reference markers (6)

Imaging Instrumentation and Measurements

In autoradiographs Feulgen-stained DNA and grain count of ^{3}H-labeled and unlabeled interphase and mitotic cells was determined with a Quantimet 720D television imaging system interfaced to a PDP-11/23 computer (4-6). Cells were selected with the light pen of the Image-Editor Module.

RESULTS

Deconvolution of the Cell Cycle Phase Distribution

The ploidy composition and size of the cycle compartments is determined separately by allowing them to empty out and by collecting their effluent for ploidy analysis. The overall frequency of ploidy subpopulations is then determined by summing their compartmental contributions.

Note the definitions "compartmental frequency" (CF), the frequency of a ploidy subpopulation within a defined compartment; "population frequency" (PF), the frequency of a subpopulation in a defined compartment with respect to the overall cell mix; "Compartmental fraction" (FR), as in S-fraction (S-FR), the fraction of the overall population resident in a specific compartment without regard to ploidy. For any subpopulation, PF = FR x CF for the specific compartment. The scheme is summarized in Table I.

S Phase. S-PF is determined in Design (I) from the DNA distribution of cells with ^{3}H-alone (interphase + mitotic) after an interval which permits complete emptying of the initial ^{3}H-labeled S compartment (Fig.1b). Interphase cells with ^{3}H-alone are in C2. These are distinguished from S cells labeled with ^{14}C. ^{3}H cells in G2+M represent the labeled S effluent. Cells with ^{3}H-alone are recored per 100 cells overall to obtain S-PF. Redistribution of the ^{3}H-cohort in G2+M At 6h and 18h is shown in Fig.1a,b. At 18h the integrated S-PF (24%) is equivalent to the S-FR, given by the pulse-labeling index. This indicates that complete emptying of S has been achieved.

G2 Phase. In Design II G2-PF is found from the DNA distribution and fraction of unlabeled mitoses scored after an interval with Colcemid (18h) that assures complete emptying of the initial G2 compartment (Fig. 1e).

G1 Phase. After pulsing (Design III) the DNA distribution of G1+G2 subpopulations is found from the DNA content of unlabeled interphase cells. (G1+G2)-PF is given by the product of the composite (G1+G2)-CF and (G1+G2)-FR. G1-PF (Fig.1h) is derived by subtracting G2-PF from the composite (G1+G2)-PF. In presenting G1-PF, G1 DNA content is doubled to correspond with the mitotic DNA ploidy index.

Mitosis. M-PF is given by the product of M-FR (2%) and M-CF (Design III). The ploidy composition of the overall population is obtained from (G1-PF + S-PF + G2-PF + M-PF), using mitotic DNA for indexing subpopulations (Fig.1k; Table I. The compartmental frequencies of subpopulations are shown in Fig. 3.

Cycle Phase distribution. This is obtained as the ratio of phase PF with respect to total cycle frequency (C-PF), Table 1, Fig. 2. Subpopulations indicated by arrows show markedly different phase distributions as illustrated in the pie charts: large G2-fraction (left). large S-fraction (center), and large G1-fraction (right). The prevalence of these patterns in the overall mix can be appreciated by reference to their plotted cycle population frequencies.

Table I: Cytokinetics of Ploidy Subpopulations

Key*	Phase	Exp	Interval	Score#	Parameters
a)	S	I	(t)	IL+ML	S efflunt
b)	S	I	(t)>Smax**	IL+ML	S effluent, S-CF, S-FR
d)	G2	II	(t)	MU	G2 effluent
e)	G2	II	(t)>G2max**	MU	G2 effluent,G2-CF,G2-FR
m)	M	III	20 min	MU	M-CF, M-FR
n)	G1+G2	III	20 min	IU	(G1+G2)-PF
h)	G1	II,III		MU,IU	G1-CF, G2-FR: (n - e)
p)	G1+G2	II	(t)	IU	residual (G1+G2)-PF
q)	G1	I,II,III	(t)		G1 effluent: (n - p -d)
j)	C	I,II,III	(t)		effluent: (a + d + q + m)
k)	C	I,II,III			C-PF: (b + e + h + m)

	Compartmental Turnover			Phase Distribution	
Key	Phase	Exp	Derivaton	Exp	Derivation
c)	S	I	a/b	I,II,III	b/k
f)	G2	II	d/e	I,II,III	e/k
	M	- - -	- - -	I,II,III	m/k
i)	G1	II,III	q/h	I,II,III	h/k
l)	C	I,II,III	j/k	I,II,III	k

#Letters correspond to Fig. 1 panels where represented.
 *IU,MU,IL,ML: unlabeled & ^{3}H-labeled interphase & mitoses.
**Smax and G2max, maximum transit-time in S and in G2.

Compartmental Turnover of Subpopulations

     Consider the estimation of 6h % compartmental turnover.
     S-Phase. It should be reiterated that the 18h labeled
S-effluent gives the magnitude of ploidy subpopulations
initially resident in S (Design I). %S turnover at 6h is
given by the ratio of the 6h and 18h labeled S effluents:
(c = a/b), Table I, Fig.1a,b,c. For the mixed population,
S-turnover was 7%/24% or 29% overall. Analysis of component
subpopulations revealed that S-turnover decreased with
increasing DNA ploidy index (Fig.1c).

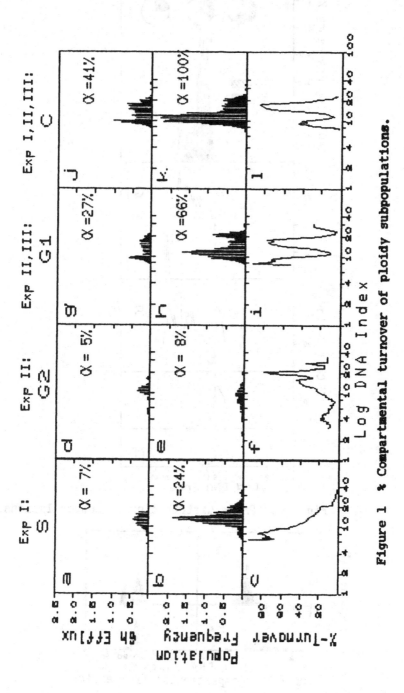

Figure 1 % Compartmental turnover of ploidy subpopulations.

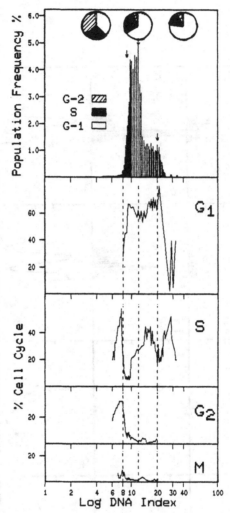

Figure 2 Phase distribution of subpopulations.

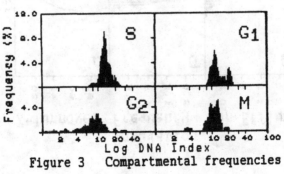

Figure 3 Compartmental frequencies

G2-Phase. The strategy is similar. In Design II, unlabeled mitoses collected by Colcemid are scored at the interval during continuous labeling. These represent the effluent from the initial G2 compartment. G2 turnover at 6h is then given by the ratio of 6h and 18h G2 effluents (f = d/e), (Fig.1d,e,f; Table I). Note that G2 turnover increased with increasing DNA ploidy index (Fig.1f). The findings suggest an increase in S-time and a decrease in G2-time with increasing ploidy.

G1-Phase. One must first determine the G1 effluent for the interval. This requires an inventory of various components (Table I). The composite (G1+G2)-PF (n) is found from unlabeled interphase cells after pulsing (Design III). Residual, unlabeled cells in G1+G2 (p) are scored after 6h continuous labeling (Design II). The combined G1+G2 effluent is then (n - p). The 6h G1 effluent (g) = (n - p - d). Note the subtraction of the 6h G2 effluent (d). %G1 turnover for subpopulations is the ratio of their 6h G1-effluent and G1-PF: i = g/h. Note two peaks of high G1 turnover (Fig.1i).

Cycle-Turnover. Total cycle effluent is obtained by summing the compartmental effluents: j = (a + d + g + m). ,Table I. % cycle turnover is the ratio of total cycle effluent and cycle PF: 1 = j/k. Two prominent subpopulations with high turnover were identified (Fig.1l).

                              DISCUSSION

    The present scheme for evaluating the cytokinetics of subpopulations exploits television imaging to automate autoradiographic analysis in conjunction with Feulgen densitometry. Algorithms have been developed for precise measurement of the Feulgen DNA content of 3H-labeled cells (4-6). The kinetic analysis is an extension of double-labeling (9) and Colcemid collection principles (2). The S-phase analysis has been reported (7,8). The overall scheme permits a deconvolution of the phase distribution of heteroploid populations and independent estimation of their compartmental turnover. The computer link provides rapid accession of multiparameter measurements and immediate graphic visualization of the cytokinetic patterns of component subpopulations. The potential of this approach lies in monitoring the cell cycle kinetics and phase distribution of subpopulations and their response to specific hormones and cytotoxins.

    The author acknowledges the excellent technical
assistance of Kenneth E. Fox. This research was generously
supported by grant 1417 from the Council for Tobacco
Research - U.S.A., Inc.

REFERENCES

1. Baisch H., et al. 1982. A comparison of mathematical
   methods for the analysis of DNA histograms by flow
   cytometry. Cell Tissue Kinet. 15:235.
2. Puck T.T. and J. Steffen. 1963. Life cycle analysis of
   mammalian cells. I. A method for localizing metabolic
   events within the life cycle and its application to the
   action of Colcemide and sublethal doses of x-irradiation.
   Biophys. J. 3:379-397.
3. Ragu U., R.J. Sklarew, J. Post and M. Levitz. 1978.
   Steroid metabolism in human breast cancer cell lines.
   Steroids 32:669-682.
4. Sklarew R.J. 1982. Simultaneous Feulgen densitometry and
   autoradiogaphic grain counting with the Quantimet 720D
   Image-Analysis System. I.Estimation of nuclear DNA content
   in ^{3}H-thymidine-labeled cells. .J Histochem. Cytochem.
   30:35-48.
5. Sklarew R.J. 1982. Simultaneous Feulgen densitometry and
   autoradiogaphic grain counting with the Quantimet 720D
   Image-Analysis System. II. Automated grain counting.
   J. Histochem. Cytochem. 30:49-57.
6. Sklarew R.J. 1983. Simultaneous Feulgen densitometry and
   autoradiogaphic grain counting with the Quantimet 720D
   Image-Analysis System. III.Improvements in Feulgen
   densitometry. J. Histochem. Cytochem. 31:1224-1232.
7. Sklarew R.J. 1984. Cytokinetics of subpopulations in
   mixed heteroploid tumors by television imaging I.
   Deconvolution of the S-phase DNA ploidy composition. II.
   Analysis of the S-phase emptying profile of ploidy
   subpopulations. J. Histochem. Cytochem. 32:413-420.
8. Sklarew R.J. 1984. Cytokinetics of subpopulations in
   mixed polyploid tumors by television imaging III.
   ^{3}H-Thymidine incorporation by ploidy subpopulations
   - Control of ^{3}H-absorbtion and relative emulsion
   efficiency in autoradiography. J. Histochem. Cytochem.
   32:421-431.
9. Wimber D.E. and H. Quastler. 1963. A ^{14}C- and 3H-thymidine
   double-labeling technique in the study of cell proliferation in Tradescantia root tips. Exp. Cell Res. 30:8-22.

CELL GROWTH, TISSUE NEOGENESIS, AND NEOPLASTIC TRANSFORMATION

Philip Skehan

The University of Calgary

Calgary, Alberta, Canada

INTRODUCTION

Exponential growth is rarely observed in vivo. most mammalian tissues, whether normal or malignant, exhibit nonexponential kinetics in which growth decelerates continuously with time (1-10). This deceleration is characterized by a gradual but progressive increase in a tissue's doubling time, and a corresponding decline in its specific growth rate.

Deceleratory growth occurs when a cell population exerts upon itself a progressive growth-inhibitory negative feedback. Growth rate is highest at a tissue's earliest point in development, and slows gradually thereafter (6,11-15). Deceleratory growth is widespread in nature, and is the single most common metazoan growth pattern (2-7, 9-18). The fine mechanisms which mediate it are not known, but the process does require some form of interaction or communication between member cells within a tissue community.

Although deceleratory kinetics cannot be modeled by exponential equations, they have been fitted with varying degrees of success to several mathematical equations, including the

Gompertz, inverse cube root (volume/surface area
ratio), logistic, and simple power functions
(1-13,15-19). With mammalian tumors, these
nonexponential deceleratory growth models have
gained considerable importance in designing
optimal schedules for adjuvant chemotherapy
(10,20), for predicting the future progress of
neoplastic disease by both primary and metastatic foci (9,10,21-23), in weighing chemother-
apeutic efficacy against host toxicity (24), and
in estimating kill fraction, possibility of
cure, and probable time to cure following
chemotherapy (24,25).

The most common mechanism of deceleration
which has been identified is the mass or self-
inhibition of growth (6,14,17,18,26-30). Mass
inhibition was first recognized in studies of
vertebrate regeneration (30), and has since been
demonstrated in a wide variety of embryonic,
postnatal, regenerative, and compensatory growth
processes at the organismic, organ, and tissue
levels (6,14,17,18,26-30). Several investigators have speculated that tumor growth might
also be governed by mass inhibition (5,8,24),
but the hypothesis has not been experimentally
tested.

Mass inhibition is a negative feedback
process in which a tissue's growth rate is
determined by its momentary size. It can be
identified experimentally by transplanting
innocula of several different sizes into host
animals. With mass inhibition, the subsequent
growth of these innocula is a function of their
momentary size only, not of their chronological
age, state of development, or innoculum size.

In principle, the regulatory mechanisms
which govern growth deceleration and mass inhibition could best be studied in cell culture
model systems. Unfortunately, mammalian cell
growth in culture is generally regarded as an
exponential process which is terminated at high
density either by medium exhaustion or postconfluency contact inhibition (18,31). Exponential and deceleratory growth are fundamentally
different kinetic processes, and one cannot be
used as a model of the other.

In this chapter I will describe several culture systems which exhibit deceleratory growth, examine their similarity to in vivo growth processes, demonstrate their governance by mass inhibition, explore the regulation of their cell cycle, and discuss the implications of mass inhibition for experimental and clinical cancer chemotherapy.

METHODS OF KINETIC GROWTH ANALYSIS

Over the years many studies of growth have appeared in the literature. With few exceptions, analysis of these data has been restricted to functions of state (size as a function of time). State analysis tells little about underlying mechanisms, and is poorly suited to extracting quantitative growth parameters of biological significance. Velocity analysis of growth, of the sort used to investigate chemical and enzymatic reactions mechanisms, has been rare despite its considerable power.

Before describing our culture models of mass inhibition and deceleratory growth, I would first like to establish the kinetic properties of the in vivo growth processes which we hope to model. A velocity analysis was performed on 58 normal and 49 neoplastic literature data bases of higher vertebrate in vivo growth processes. The data bases examined (Table 1) presented measurements of tissue size as a function of time. From these we calculated the specific growth rate (SGR) and doubling time (Td) for each successive pair of data points:

$$SGR = 100(-1 + 2^{Tu(Ln[S2/S1]/(0.69315)Tobs)})$$

$$Td = 16.636/Ln(1 + 0.01SGR)$$

where S1 and S2 are successive size measurements separated from one another by a period of time Tobs (in days); Tu is the unit period of time (1 day), SGR is in units of percent increase in size per day, and Td is measured in hours. Three growth phases can be identified from such an anaysis: acceleratory (increasing SGR); exponential (constant SGR); and deceleratory.

TABLE 1. IN VIVO DATA BASES.

NORMAL DATA BASES		NUMBER		NUMBER
MAMMALIAN:	PIG	19	CAT	2
	RAT	15	SHEEP	2
	MOUSE	5	HUMAN	1
			PIKA	1
AVIAN:	FIELDFARE	6	GOOSE	1
	JACKDAW	3	QUAIL	1
	CHICKEN	2		
POSTNATAL		48	INTERNAL ORGANS	30
EMBRYONIC/POSTNATAL		6	WHOLE ANIMAL	28
EMBRYONIC		4		

TUMOR DATA BASES:			
MOUSE	34	CARCINOMAS	24
RAT	14	SARCOMAS	11
HUMAN	1	GLIOMAS	2
		MELANOMAS	2
LEUKEMIA/LYMPHOMAS	2	UNCERTAIN	7
PLASMOCYTOMAS	1		

(decreasing SGR).
     The ability of 18 different growth
equations to model the data bases was examined
(Table 2). Each of the equations was linearly
transformable. We therefore employed linearly
transformed data to permit a linear regression
analysis of all models. For each equation, the
least squares correlation coefficient (Rc) and a
nonparametric index of curvature (C) were
calculated. Like Rc, C ranges in value from 0
to 1. A value of 0 indicates linearity, while a
value of 1 indicates extreme curvature. The C
test is based upon the principle that if data
conformed well to the linearity hypothesis and
deviated from the least squares regression line
only by random error normally distributed around
expected values, then there should be on average
as many data points above the regression line as
below it. By contrast, where transformed data
deviated from the linearity hypothesis, some
regions would possess an excess of points above
the linear regression line and other regions

TABLE 2.   MATHEMATICAL GROWTH MODELS.

#		FAMILY OF EQNS	EQUATION
1.	N = 1	Nth ROOT	$SGR = R(1 - S^N/K^N)$
2.	1/2		
3.	1/3		
4.	1/4		
5.	N = -1	INVERSE Nth ROOT	$SGR = \theta(1 - K^{-N}/S^N)$
6.	-1/2		
7.	-1/3		
8.	-1/4		
9.	N = 2	Nth POWER	$SGR = R(1 - S^N/K^N)$
10.	3		
11.	4		
12.	N = -2	INVERSE Nth POWER	$SGR = \theta(1 - K^{-N}/S^N)$
13.	-3		
14.	-4		
15.	GOMPERTZ		$SGR = G(\text{Ln } K - \text{Ln } S)$
16.	EXPONENTIAL DECAY		$SGR = Re^{-GS}$
17.	HYPERBOLIC		$SGR^{1/2} = G(1 - [\text{Ln } S/\text{Ln } K])$
18.	SIMPLE POWER		$\text{Ln } SGR = \text{Ln } G - (1/b)\text{Ln } S$

SGR = specific growth rate.
S = tissue or tumor size.
K = final size which tissue or tumor reaches.
R = specific growth rate at infinitesimal size.
θ = specific growth rate at infinite size.
G,b = arbitrary rate coefficients with no obvious biological meaning.

would possess an excess of points below it.
Graphical inspection of the data bases showed
that the curvature of SGR-size plots was
restricted to simple downward convexity and
concavity.   Thus curvature could be examined by
dividing each data plot into just 3 regions -
left (l), middle (m), and right (r).   C was
calculated from the equation:

$$C = \frac{1}{3}\left(\frac{El}{0.5\ Nl} + \frac{Em}{0.5\ Nm} + \frac{Er}{0.5\ Nr}\right) = \frac{2}{3}\sum\frac{Ei}{Ni}$$

where Ni is the number of datas points in a
region, 3 is the number of regions, and Ei is
the excess of points above 0.5 Ni either above
or below the regression line in a particular
region.   A goodness-of-fit index (G) was
calculated as the average of Rc and (1-C).
G gives equal weighting to correlation

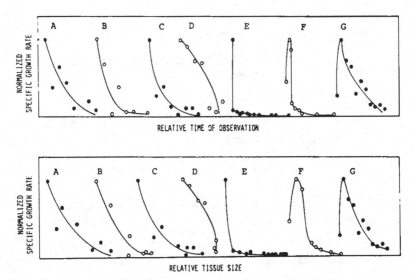

Fig. 1. Representative normal tissue data bases.
From right to left in each panel: swine
gastrocnemius; swine adrenals; swine tibia;
swine eyes; strong wool ram; fine wool ram; rat
body weight (43-45).

coefficient and linearity as determinants of an
equation's ability to model a data base. High
values of G reflect a good fit while low values
indicate a poor fit.

NORMAL PATTERNS OF GROWTH.

Normal tissues exhibited three kinetic
patterns of growth (Fig. 1, Table 3). Pure
deceleratory growth was the most common. Growth
was fastest at the earliest developmental age
examined, and slowed gradually but progressively
thereafter. A less common pattern consisted of
a brief acceleratory phase followed by a much
longer period of deceleratory growth. The least
common normal growth pattern consisted of a
prolonged deceleratory phase ending in a period
of limited tissue regression which presumably
reflects compensation for an overshoot of the
tissue's set point for final adult size. None
of the data bases possessed an exponential
phase. Exponential growth in nature appears to

TABLE 3. KINETIC PATTERNS OF IN VIVO GROWTH.

	D	AD	ED	DED	KD	DR
NORMAL DATA BASES						
NUMBER	49	7	0	0	0	2
PERCENT	84	12	0	0	0	3
TUMOR DATA BASES						
NUMBER	34	10	2	2	1	0
PERCENT	69	20	4	4	2	0

TABLE 4. CURVATURE OF SGR-SIZE PLOTS.

	DOWNWARD CONVEXITY	LINEARITY	DOWNWARD CONCAVITY
NORMAL DATA BASES	53	4	1
TUMOR DATA BASES	47	1	0

be uncommon. Regardless of the particular
growth pattern a tissue exhibited, the
deceleratory was consistently the most prominent
phase of growth.

The normal data bases did not permit the
mass inhibition hypothesis to be tested. Data
were replotted as SGR versus time (Fig. 1),
however, to at least examine its plausibility.
In all of the data bases, SGR decreased
monotonically with size suggesting that mass
inhibition might be operating. Since there is
no evidence that the growth deceleration of
normal tissues results from the development of
vascular insufficiency, and since the
deceleratory phase generally extends from early
embryogenesis into early adulthood, it is likely
that deceleration results from mass inhibition
in many of the data bases.

Of the 58 normal data bases, 53 had a
deceleratory phase that exhibited downward
convexity when SGR was plotted against size
(Table 4). Downward convexity indicates that a
unit increase in size exerts a stronger growth
inhibitory effect on a small tissue than on a
large one. Its practical implication is that
most of the total growth inhibition to which a
tissue is eventually subjected develops early in
a tissue's growth while it is still quite small
in size. Opposite curvature (downward concavity) indicates that little growth inhibition

TABLE 5. GOODNESS-OF-FITS TO GROWTH EQUATIONS.

EQN #			NORMAL DATA BASES		TUMORS	
			MEAN G	# BEST FITS	MEAN G	# BEST FITS
1	Nth ROOT	N=1	.657	3	.560	1
2		N=.5	.693	3	.612	1
3		N=.333	.709	2	.629	0
4		N=.25	.719	1	.634	2
5	INVERSE	N=-1	.741	9	.713	5
6	Nth ROOT	N=-.5	.774	4	.717	6
7		N=-.333	.782	2	.718	3
8		N=-.25	.787	7	.714	3
9	Nth POWER	N=2	.627	1	.502	0
10		N=3	.476	0	.365	0
11		N=4	.572	0	.444	0
12	INVERSE	N=-2	.676	1	.685	4
13	Nth POWER	N=-3	.655	4	.669	1
14		N=-4	.640	1	.652	5
15	GOMPERTZ		.755	2	.693	2
16	EXP. DECAY		.724	4	.617	4
17	HYPERBOLIC		.785	9	.725	6
18	SIMPLE POWER		.727	5	.711	6

occurs until a tissue is quite large and approaching its final adult size. This pattern of growth was observed in only 1 of the 58 normal data bases. A linear decay of growth rate with size was observed in 4 cases.

Table 5 summarizes the goodness-of-fit analysis for the normal tissue deceleratory phases. Two conclusions are immediately obvious. First, there is no single equation which consistently provides the best model of normal growth. Second, of the commonly used growth equations only the inverse cube root (volume/area ratio) was well above average as a general in vivo growth model. 16 of the 18 growth equations provided at least 1 best fit to the normal data bases. The inverse 4th root equation (N=-.25) provided the highest mean value of G, the Spillman (N=-1) and hyperbolic equations had the most best-fits, the inverse

TABLE 6. NORMAL GROWTH CHARACTERISTICS.

1. GROWTH IS EXCLUSIVELY OR PREDOMINANTLY DECELERATORY.
2. DECELERATION IS PROBABLY CAUSED BY MASS INHIBITION IN MANY INSTANCEs.
3. MOST OF A TISSUE'S GROWTH INHIBITION IS IMPOSED EARLY IN ITS DEVELOPMENT WHEN IT IS STILL QUITE SMALL (DOWNWARD CONVEXITY OF SGR-SIZE PLOTS).
4. INVERSE Nth ROOT EQUATIONS PROVIDE THE BEST FAMILY OF GROWTH MODELS, Nth POWER EQUATIONS THE WORST.

Nth root was the best and the Nth power the worst family of growth equations (Table 6).

TUMOR GROWTH PATTERNS.

Tumor growth was also predominantly or exclusively deceleratory (Fig. 2). Of the 49 tumor data bases, all contained a deceleratory phase (Table 3). 34 exhibited pure deceleratory (D) kinetics at all observed times, 10 had an acceleratory- deceleratory (AD) pattern, one exhibited an initial kill-off followed by deceleratory growth (KD), two began with a brief exponential phase that was followed by a deceleratory phase (ED), and two had a brief exponential phase in the middle of an otherwise deceleratory pattern (DED). These last 2 exponential phases are probably artifacts of data fluctuation. In any event, the predominant mode of growth for tumors, as for normal tissues, was deceleratory.

Seven of the tumor data bases provided growth data for innocula of more than one size, and therefore permitted a test of the mass inhibition hypothesis. Of the 7, 4 exhibited classical mass inhibition (Figs. 3,4) and 1 did not; the remaining 2 were equally compatible with both mass inhibition and age dependent growth regulation, so that no conclusion could be made about the nature of their control. Thus the growth of at least 4 and possibly 6 of the 7

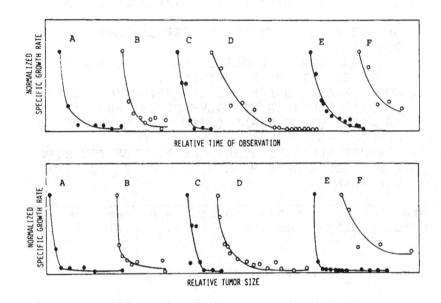

Fig. 2. Representative tumor data bases. From
left to right in each panel: mouse MTG-B mammary
carcinoma; rat R4A2 sarcoma; mouse L1210
ascites; mouse P1798 lymphosarcoma; rat lewis
lung carcinoma; rat R3A45 (3,24,46-48).

---

tumors was governed by mass inhibition.

As with normal tissues, the great majority
of tumor deceleratory phases exhibited downward
convexity in their SGR-size plots (Fig. 2, Table
4). Most of the growth inhibition to which the
tumors were ultimately subjected was imposed
early in their development while they were still
small.

Again as with normal tissues, no single
growth equation best modeled all data bases
(Table 5). 14 different equations provided at
least 1 best fit to the tumor data, the inverse
square root equation had the highest G value,
the hyperbolic and inverse square root equations
gave the most best-fits, the inverse Nth root
was the best family of growth equations, and the
inverse Nth power the was the worst.

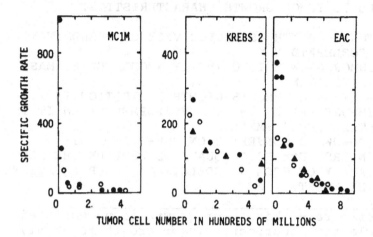

Fig. 3. Mass inhibition of ascites tumor growth in vivo. Calculated from data of refs. 4, 5, 8.

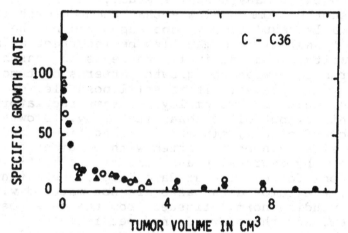

Fig. 4. Mass inhibition of solid tumor growth in vivo. Calculated from data of ref. 49.

COMPARISON OF NORMAL AND NEOPLASTIC GROWTH.

The predominant characteristics of in vivo tumor growth are summarized in Table 7. They are effectively identical to those of normal growth processes (Table 6).

This unexpected finding is at odds with a number of popular views about neoplastic growth. The statement so commonly encountered in the

TABLE 7. TUMOR GROWTH CHARACTERISTICS.

1. TUMOR GROWTH IS EXCLUSIVELY OR PREDOMINANTLY
   DECELERATORY.
2. TUMOR GROWTH IS COMMONLY REGULATED BY MASS
   INHIBITION.
3. MOST OF A TUMOR'S GROWTH INHIBITION IS
   IMPOSED EARLY IN ITS DEVELOPMENT WHEN IT IS
   STILL QUITE SMALL
   (DOWNWARD CONVEXITY ON SGR-SIZE PLOTS).
4. INVERSE Nth ROOT EQUATIONS PROVIDE THE BEST
   FAMILY OF GROWTH MODELS, Nth POWER EQUATIONS
   THE WORST.

---

literature - that tumor growth is unregulated,
is clearly incorrect. Tumor growth is highly
regulated, and the underlying control mechanism
of mass inhibition is the same one which governs
the growth of many normal tissues.
     Another common view - that tumor growth is
abnormally regulated, is not supported by the
present analysis and may also be incorrect as a
generality. Qualitatively, there is no sensible
difference between the growth patterns of tumors
and normal tissues. It is still possible, of
course, that the two employ the same regulatory
mechanisms, but differ quantitatively. Growth
rate coefficients, minimum doubling times, and
final sizes can be extracted with considerable
accuracy by regression analysis of the best-fit
equations for particular data bases. Unfortunately, the tumor data bases are not matched with
corresponding normal tissues from the same host
species, and therefore do not permit
quantitative comparison.   However, Steel (15)
has compiled an exhaustive tabulation of
tritiated thymidine data on this question.
While his compilation does show that certain
kinds of tumors tend to have shorter doubling
times than their normal tissues of origin, it
shows that certain other tumors tend to have
slower and still others have the same doubling
time as their normal counterparts. On balance,
there is no sensible pattern of difference in
doubling time between tumors as a group and
normal tissues as a group. Thus the data which

are currently available do not support the
hypothesis that tumor cell growth is generally
more rapid than that of normal cells. While the
hypothesis is almost certainly correct for
certain types of tumors, it appears to be
incorrect as a general proposition. It follows
that if most normal and neoplastic growth
patterns are both qualitatively and
quantitatively identical, then most cancers
cannot be diseases of abnormal growth.

CANCER AS TISSUE NEOGENESIS.

     If cell growth is not the critical lesion
in many cancers, what then is? The phenomenon
of mass inhibition offers a possible answer.
Mass inhibition results from a growth inhibitory
negative feedback communication between member
cells within a multicellular community. Three
types of signals are likely to mediate this
communication process. Direct cell contact
interactions have been shown to do so in at
least one tumor cell culture system, diffusable
growth inhibitors are implicated in several in
vivo systems, and the extracellular matrix may
contribute in some cell culture systems (23,
32-35). These interactions all tend to have
some degree of target specificity, and therefore
effectively constitute cellular recognition
mechanisms.
     This suggests that deceleratory growth
caused by mass inhibition may involve two
separate steps: (1) an initial cellular
recognition event which, if effective, acts as a
switch to activate mass inhibition; and (2) the
actual mass inhibition per se. If this
hypothesis is correct, then mass inhibition
could only occur if an appropriate recognition
has first been established.
     This suggests that neoplastic
transformation may arise from an alteration in
the cellular recognition processes which mediate
mass inhibition. Tumors behave as if they were
new types of tissues with normal growth regulatory policies and control mechanisms but

altered or inappropriate recognitive
determinant. Thus neoplastic transformation
may, in some cases, be a disease of tissue
neogenesis.

In a mass inhibited tissue, growth is
fastest at infinitesimal size. As size
increases, the rate of growth slows. Initiated
cells during the earliest stages of preneoplastic phenotypic progression within a mature
adult tissue respond normally to the tissue's
growth regulatory signals (36-38). As
preneoplastic progression continues and the
initiated cell gradually acquires additional
phenotypic alterations, all that is necessary
for its neoplastic conversion is a change in the
nature of its growth regulatory recognitive
determinants. Such a change would make it
unresponsive to the mass inhibitory signals of
the surrounding normal cells. The transformed
cell would now have unique recognitive
determinants. Having no access to other cells
with the same or complimentary determinants, it
would suddenly be released from mass inhibition
and would undergo an explosive acceleration in
its growth. As growth enlarged the tumor cell
population, mass inhibition would gradually
develop and the tumor's growth rate would begin
to slow, giving rise to deceleratory kinetics.
Growth would eventually stop altogether when the
tumor reached its equivalent of a mature adult
size, provided of course that it did not first
kill its host. What causes such a tumor to
temporarily grow faster than its normal tissue
counterpart is not a change in either its growth
or growth regulatory machinery per se, but
rather the sudden loss of access to cells with
identical or complimentary recognitive
determinants.

The ostensibly faster growth of tumors, so
widely asserted in the literature yet
contraditcted by available cell cycle data, is
actually an artifact of analysis. In deceleratory growth, there is no single characteristic
value of growth rate or doubling time. Both
assume an infinite number of different values
during the overall deceleratory process. To

make valid quantitative comparisons between
normal tissue and tumor growth characteristics,
it is necessary to compare them at the same
specified size(s). There is no evidence that
the kinetic parameters of normal tissue and
tumor growth show any consistent differences
when sizes are comparable. Tumors wrongly
appear to grow faster than normal tissues only
when relatively small tumors are incorrectly
compared with large and fully mature adult
tissues. This is not a valid comparison. In
the tissue neogenesis hypothesis which I have
outlined, a tumor may sometimes experience a
change in its growth machinery so that it will
grow either faster or slower that its normal
counterpart at a specific size, but this change
is incidental to neoplastic transformation and
is not required for progressive neoplastic
growth.

IMPLICATIONS FOR CLINICAL AND EXPERIMENTAL CHEMOTHERAPY.

Cancer chemotherapy is presently confronted
by two dilemmas: the antitumor drugs that are
now available (largely antiproliferatives) work
poorly if at all against most forms of cancer;
and the animal and culture model systems that we
use to develop new drugs have little predictive
value for clinical use. Why the antiproliferatives are so effective in models yet have such
limited clinical effectiveness, is among the
most important unsolved problems in cancer
research.

The deceleratory growth kinetics of tumors,
combined with the downward convexity of most
SGR-size plots, argues strongly that the
antiproliferatives will not be effective except
at small tumor size when rate of growth is high.
By the time that tumors reach clinically
detectable size, they are already growing very
slowly in most cases, and can therefore be
expected to respond poorly to antiproliferative
chemotherapy. Indeed, because of their
considerable toxicity to many host tissues, the

antiproliferatives may actually prove counter-
productive as a primary therapy against large
tumor masses.

Tumor sensitivity to antiproliferative
chemotherapy can only be expected when tumors
are very small, generally much below the minimum
clinically detectable size. This situation only
occurs during the earliest stages of neoplastic
disease (when tumors are usually undetectable)
or following the therapeutic elimination of most
tumor burden. This latter consideration
provides a powerful argument for the use of
adjuvant antiproliferative chemotherapy against
small residual tumor masses following the
resection or irradiation of bulk mass.

The in vivo analysis also provides an
explanation for the commonplace failure of
experimental antiproliferative chemotherapy
models to accurately predict clinical drug
efficacy. In cell culture models, drug efficacy
is almost invariably tested under conditions in
which the target cells are rapidly growing and
therefore sensitive to antiproliferatives. By
contrast, clinical tumors are usually in an
advanced state of growth deceleration by the
time they are detected, and are therefore not
likely to respond well to antiproliferative
therapy. A similar consideration probably
explains the poor predictve ability of in vivo
animal models as well. High density, slowly
growing mass inhibited cultures and comparativ-
ely large tumors selected for rapid decay of
growth rate with size are likely to prove more
reliable indicators of clinical drug efficacy
than the models currently in use.

MASS INHIBITION IN CULTURE.

We have identified 3 cell lines which in
culture exhibit a prominent deceleratory phase
that is mediated by mass inhibition. These are
the rat C6 glioma, mouse SVS, and canine MDCK
lines. The growth characteristics of the 3
lines are nearly identical, and differ only in
minor ways. A detailed characterization of C6

growth has already been published (31,40,41),
and will be presented here in summary form only.
    When cells are subcultured, they proceed
through a culture growth cycle consisting of 3
successive phases: (1) lag; (2) acceleratory;
and (3) deceleratory. The first two phases are
brief, so that deceleration is the predominant
mode of growth. During deceleration, growth
exhibits the classical characteristics of a mass
inhibited process. The specific growth rate is
a monotonically decreasing function of momentary
population density, but is independent of both
initial density and culture age. This mass
inhibition correlates with the degree of contact
cells make with their neighbors, but does not
involve growth inhibitory conditioned medium
factors, the extracellular matrix, or medium
depletion effects. SGR-density curves exhibit
downward convexity that is mild for C6 cells,
moderate for SVS, and strong for MDCK. In all
cases, the inverse Nth root family of equations
provided the best models of the deceleratory
phase. Thus kinetically the deceleratory phase
of these cell culture lines closely parallels
that of in vivo tumors and tissues.
    The contact interactions which mediate mass
inhibition are very different from those
postulated by conventional postconfluency
contact inhibition theory. We have termed the
mass inhibition process contact modulation to
avoid confusion. In contact modulation a cell's
growth rate is a monotonically decreasing
function of the percentage of its total cell
surface area that is in contact with other
surfaces. Any change in the extent of contact,
no matter how small, causes a compensatory
change of opposite direction in the specific
growth rate, and the change in growth rate is
proportional to the change in contact area.
Contact modulation therefore acts at all
densities in which even minimal cell contact
occurs. For most types of cells this range
encompasses the entire spectrum of culture
operating densities from very sparse
subconfluency to heavily multilayered
supraconfluency. Confluency per se is

irrelevant to the process. Contact modulation
provides a simple explanation of the
acceleratory-deceleratory growth pattern which
follows the inertial lag phase. The reduction
of contact by subculture partly releases cells
from mass inhibition, and growth acceleration
results. This acceleration continues until
cells achieve the new and higher growth rate
appropriate for their new but reduced extent of
contact. At this point acceleration ends, and
cells are growing at the fastest rate they will
achieve in this particular culture growth cycle.
Continued growth increases population density
and therefore cell contact, gradually reimposing
mass inhibition, which is the cause of the
deceleratory phase.

Although mass inhibition is contact
mediated in the cell lines we have thus far
examined, the concept can be generalized to
include any intercellular communication
mechanism whose growth inhibitory effectiveness
is proportional to tissue size or population
density but independent of initial size/density
and chronological time. In other words, this
same concept can, when generalized, accomodate
diffusable growth inhibitors and the extracellular matrix as well as cell contact
interactions.

CELL CYCLE ANALYSIS OF DECELERATORY
GROWTH IN VITRO.

Control of the C6 cell cycle during contact
modulated mass inhibition was examined by
Acriflavin-Feulgen microspectrophotometry of
nuclear DNA contents. Cell cultures were plated
at several seeding densities, and samples
collected on successive days throughout both the
acceleratory and deceleratory phases. Median G2
content together with upper and lower G2 bounds
were determined on cells metaphase-arrested by
colcemid, and the values were halved to obtain
corresponding G1 parameters. Because S phase
was large by comparison with G1 and G2, and
generally asymmetric in shape, ordinary flow

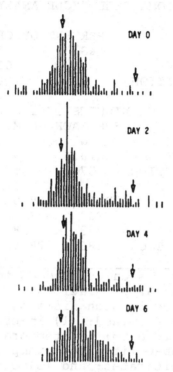

Fig. 5. Histograms of cellular DNA content.
Frequency of cells (ordinate) is plotted against
DNA content channel (abscissa) for cells
collected over a 6 day period. The day 0 cells
were taken from a heavily multilayered,
medium-depleted culture and show a prominant G1
accumulation. Doubling times on days 0, 2, 4,
and 6 were 113, 23.8, 73.5, and 563 hours
respectively.

cytometry-type cell cycle curve-fitting
algorithms could not be used for phase analysis.
Instead, we calculated pseudophase values which
approximated G2 as twice the number of cells
with DNA content greater than or equal to median
G2 in the colcemid-blocked sample, and G1 as
equal to twice the number of cells with DNA
content less than or equal to median G1. These
criteria were used because overlap by S phase
should be negligible in the lower half of the G1
and the upper half of the G2 DNA channels. S

TABLE 8.   C6 GLIOMA CELL CYCLE ANALYSIS.

a.   PERCENT OF CELLS IN PHASE

	%G1	%S	%G2
MASS INHIBITION	20	67	14
DEPLETION INHIBITION	72	25	3

b.   MINUTES THAT A PHASE CHANGES
                                FOR EACH 60 MIN. Td CHANGE

	G1	S	G2
MASS INHIBITION	10.6	46.8	2.6
DEPLETION INHIBITION	57.2	11.6	-9.2

c.   PROPORTIONAL PHASE INCREASE (%)

	Td	G1	S	G2
MASS INHIBITION	293	328	328	142
DEPLETION INHIBITION	386	2804	236	-75

phase frequency was calculated as total cells
minus the G1 and G2 populations.

Five cell cycle mechanisms of C6 mass
inhibition were examined: (1) in-cycle arrest;
(2) cycle exit into Go; (3) temporary arrest
(transition probability, for example); (4)
quantized transit rates: and (5) continual
cycling with slowing of transit progression.
The first four mechanisms all lead to a
prominent accumulation of cells at specific
restriction points in the cell cycle, while the
last mechanism does not.

Two separate mechanisms of cell cycle
inhibitory control were observed (Fig. 5). The
first operated at all densities and resulted
from mass inhibition. It involved a gradual
slowing of the cycle transit rate without any
detectable arrest. The second was a G1
accumulation, which in well fed cultu es did
not occur until extremely high densities of 4-5
times confluency (4-5 C) were reached. It was
in this same density range that medium depletion
effects began to develop. The G1 arrest
therefore appears to reflect the development of
either a nutrient or serum factor deficiency.
The G1 accumulation had the characteristics of a
Lajtha cycle-exit process rather than a Gelfant
in-cycle arrest (44,45): following release from

arrest, a number of hours were required before growth accelerated.

In cultures that were strongly depletion inhibited (density circa 5C), cells accumulated primarily in G1. S phase was small and G2 was negligible (Table 8a). When these cells were subcultured and allowed to reach a mass inhibited steady state withour detectable depletion effects, S became the predominant phase, and G1 and G2 were both moderately small. In depletion inhibited cultures, G1 changed by an average of 57.2 minutes for each 60 minute change in doubling time (Table 8b). By contrast, during steady state mass inhibition a change in S phase accounted for 46.8 minutes of each 60 minute Td change.

When the proportional change in phase duration is considered instead of the absolute, it can be seen that G1 changes disproportionately in depletion inhibited cultures, while in mass inhibition each phase changes to a roughly similar degree (Table 8c).

ACKNOWLEDGEMENTS.

I would like to thank Ms. Ping Chiu, Mr. Stanley Cheng, Dr. Susan Friedman, and Dr. James Thomas for their many contributions to this project, which was supported by the Alberta Heritage Fund for Applied Cancer Research, The Alberta Heritage Fund for Medical Research, and The Medical Research Council of Canada.

REFERENCES

1. Mottram, J.C. and S. Russ. 1917-1918.
   Proc. Roy. Soc. London Ser. B 90:1.
2. Mayneord, W.V. 1932. Am. J. Cancer 16:841.
3. Schrek, R. Am. J. Cancer 28:345.
4. Klein, G. and L.Revesz. 1953. J. Natl.
   Cancer Inst. 14:229.
5. Patt, H.M. and M.E. Blackford. 1954. Cancer
   Res. 14:391.
6. Bertallanfy, von L. 1957. Quart. Rev. Biol.
   32:217.
7. McCredie, J.A., W.R. Inch, J. Kruv, and T.A.
   Watson. Growth 29:331.
8. Lala, P.K. and H.M. Patt. 1966. Proc. Natl.
   Acad. Sci. 56:1735.
9. Laird, A.K. 1969. Natl. Cancer Inst. Monogr.
   #30:15.
10. Simpson-Herren, L. and H.H. Lloyd. 1970.
    Cancer Chemotherapy Repts. 54:143.
11. Pearl, R. 1925. The Biology of Population
    Growth. Knopf, New York.
12. Winsor, C.P. 1932. Proc. Natl. Acad. Sci.
    18:1.
13. Smith, F. 1952. Ecology 33:441.
14. Rose, S.M. 1957. Biol. Rev. 32:331.
15. Steel, G.G. 1977. Growth Kinetics of Tumors.
    Clarendon Press, Oxford.
16. Laird, A.K. 1965. Growth 29:249.
17. Stebbing, A.R.D. 1981. J. Mar. Biol. Assoc.
    U.K. 61:35.
18. Skehan, P. and S.J. Friedman. 1984. Cell
    Tissue Kinet., in press.
19. Dethlefsen, L.A., J.M.S. Prewitt, and M.L.
    Mendelsohn. 1968. J. Natl. Cancer Inst.
    40:389.
20. Vaage, T. and M. Costanza. 1979. Cancer Res.
    39:4466.
21. Durbin, P, N. Jeung, M. Williams, and J.
    Arnold. 1967. Cancer Res. 27:1341.
22. Simpson-Herren, L., T. Springer, A. Sanford,
    and J. Holmquist. 1977. Progr. Cancer Res.
    Therapy 5:117.
23. Sugarbaker, E.V., J. Thornthwaite, and A.S.
    Ketcham. 1977. Progr. Cancer Res. Therapy
    5:227.

24. DeWys, W.D. 1972. Cancer Res. 32:374.
25. Cox, E.B. 1978. Proc. Am. Assoc. Cancer Res. 19:184.
26. Weiss, P. and J.L. Kavanau. 1957. J. Gen. Physiol. 41:1.
27. Tanner, J.M. 1981. Brt. Med. Bull. 37:233.
28. Goss, R.J. 1969. In R.J. Goss (ed.) "Regulation of Organ and Tissue Growth," pp. 1-11, Academic Press, New York.
29. Snow, M.H.L. 1981. Brt. Med. Bull. 37:221.
30. Ellis, M.M. 1909. J. Exp. Zool. 7:421.
31. Skehan, P. and S.J. Friedman. 1982. Cancer Res. 42:1636.
32. Schatten, W.E. 1958. Cancer 11:455.
33. Ketcham, A.S., D.L. Kinsey, H. Wexler, and N. Mantel. 1961. Cancer 14:875.
34. Gorelik, E., S. Segal, and M. Feldman. 1981. Int. J. Cancer 27:847.
35. Weiss, L., G. Postc, A. MacKearnin, and K. Willett. 1975. J. Cell Biol. 64:135.
36. Foulds, L. 1969. "Neoplastic Development." Academic Press, New York.
37. Emmelot,P. and E. Scherer. 1980. Biochim. Biophys. Acta 605:247.
38. Farber,E. and R. Cameron. 1980. Adv. Cancer Res. 31:125.
39. Skehan, P. 1976. Exp. Cell Res. 97:184.
40. Skehan, P. and S.J. Friedman. 1976. Exp. Cell Res. 101:315.
41. Lajtha, L. 1963. J. Cell. Comp. Physiol. 62 (Suppl. 1):143.
42. Gelfant, S. 1977. Cancer Res. 37:3845.
43. Tumbleson, M.E., O.W. Tinsley, J.B. Mulder, and R.E. Flatt. 1970. Growth 34:401.
44. McCarthy, P.H. and Butterfield, R.M. 1981. Growth 45:351.
45. Clark, R.G. and M.F. Tarttelin. 1978. Growth 42:113.
46. Clifton, K.H. and Yatvib, M.B. 1970. Cancer Res. 30:658.
47. Dombernowsky, P. and N.R. Hartmann. 1972. Cancer Res. 32:2452.
48. Davis, J.M., A.K. Chan, and E.A. Thompson. 1980. J. Natl. Cancer Inst. 64:55.
49. Sato, N., M. Michaelides, and M.K. Wallack. 1981. Cancer Res. 41:2267.

INDEX